KU-260-803

Practical Problems in Clinical Psychiatry

Edited by

KEITH HAWTON
Consultant Psychiatrist and Clinical Lecturer,
University Department of Psychiatry and
Warneford Hospital, Oxford

and

PHILIP COWEN
MRC Clinical Scientist
Medical Research Council Clinical Pharmacology Unit,
Littlemore Hospital, Oxford

OXFORD   NEW YORK   TOKYO
OXFORD UNIVERSITY PRESS
1992

Oxford University Press, Walton Street, Oxford OX2 6DP
Oxford New York Toronto
Delhi Bombay Calcutta Madras Karachi
Kuala Lumpur Singapore Hong Kong Tokyo
Nairobi Dar es Salaam Cape Town
Melbourne Auckland Madrid
and associated companies in
Berlin Ibadan

Oxford is a trade mark of Oxford University Press

Published in the United States
by Oxford University Press Inc., New York

© K. Hawton and P. Cowen, 1992

All rights reserved. No part of this publication may be
reproduced, stored in a retrieval system, or transmitted, in any form
or by any means, without the prior permission in writing of Oxford
University Press. Within the UK, exceptions are allowed in respect of any
fair dealing for the purpose of research or private study, or criticism or
review, as permitted under the Copyright, Designs and Patents Act, 1988, or
in the case of reprographic reproduction in accordance with the terms of
licences issued by the Copyright Licensing Agency. Enquiries concerning
reproduction outside those terms and in other countries should be sent to
the Rights Department, Oxford University Press, at the address above.

This book is sold subject to the condition that it shall not,
by way of trade or otherwise, be lent, re-sold, hired out, or otherwise
circulated without the publisher's prior consent in any form of binding
or cover other than that in which it is published and without a similar
condition including this condition being imposed
on the subsequent purchaser.

A catalogue record for this book is available from the British Library

Library of Congress Cataloging in Publication Data
Practical problems in clinical psychiatry
edited by Keith Hawton and Philip Cowen.
(Oxford medical publications)
Companion v. to: Dilemmas and difficulties in the management of
psychiatric patients / edited by Keith Hawton and Philip Cowen. 1990.
Includes bibliographical references.
1. Mental illness—Treatment—Congresses. I. Hawton, Keith,
1942– . II. Cowen, Philip. III. Title: Dilemmas and difficulties
in the management of psychiatric patients. IV. Series.
[DNLM: 1. Mental Disorders—therapy. WM 400 P8945]
RC475.5.P68 1992   92-49369
ISBN 0-19-262334-6 (hbk.)
ISBN 0-19-262333-8 (pbk.)

Typeset by Downdell, Oxford
Printed in Great Britain by
Bookcraft (Bath) Ltd
Midsomer Norton, Avon

RC
475
.S
.PRA

OXFORD MEDICAL PUBLICATIONS

Practical Problems in Clinical Psychiatry

Preface

Many practical problems arise during clinical management in psychiatry which pose difficulties for both clinicians, be they psychiatrists, general practitioners, physicians, or other health professionals, and their patients. In part this is because of our lack of specific knowledge of the aetiology of so many psychiatric disorders. In addition the scientific evaluation of treatment strategies, an area of increasing contemporary interest, is beset with methodological pitfalls. For many of the clinical problems that confront us there are no straightforward, easily applicable solutions and we must therefore rely on what we can glean from the scientific literature together with the advice of more experienced colleagues.

It was with these clinical problems in mind that the present book was planned. We wanted to explore in a very practical way issues that frequently cause difficulties in management for both trainees and experienced psychiatrists, as well as for other clinicians. Accordingly, the volume covers a wide but, we think, important and relevant range of topics, including psychiatric management in general practice, the indications for new antidepressants, the role of psychotherapy in manic-depression and personality disorder, through to current treatment issues in liaison psychiatry, sexual dysfunction, anorexia nervosa, and schizophrenia. Also addressed are issues which are becoming increasingly relevant, such as how to care for doctors who have emotional problems and how to help with the psychological and psychiatric difficulties faced by many survivors of disasters.

We invited practitioners with long-standing clinical and research interests in the particular topics to provide initial articles which were then each discussed by all the contributors at a two day meeting at Green College, Oxford, in April 1991. In this way, a wealth of clinical experience and knowledge was brought to bear on all of the areas covered which facilitated the subsequent preparation of the chapters included in this book. We have previously found this format particularly appropriate for the construction of chapters that provide practical clinical advice and there have been favourable reviews of an earlier volume prepared in this way (*Dilemmas and difficulties in the management of psychiatric patients*, 1990, edited by K. Hawton and P. Cowen, Oxford University Press). Three further chapters were obtained from authors who were unable to attend the meeting in Oxford.

We would like to thank Dista Products Ltd, Duphar Laboratories Ltd, E. Merk Ltd, Pfizer Ltd, and Upjohn Ltd whose generous support made the meeting of contributors at Green College possible. The meeting itself proved to be a great success and we believe that the quality of the practical advice contained in each chapter was greatly enhanced by the opportunity of the contributors to compare research findings and clinical experience.

The management of psychiatric disorders is not an easy task but we hope that the practical strategies outlined in these chapters will be helpful to all clinicians involved in this area. In particular, we believe that the clinical advice this book contains will be of great value to trainees who often, in the early stages of their career, are faced with issues of patient management that their seniors would find equally taxing.

Oxford   K.H.
November 1992   P.C.

Contents

Contributors

Christopher Bass
Consultant in Liaison Psychiatry, John Radcliffe Hospital, Oxford, UK

Patricia R. Casey
Professor of Psychiatry, Mater Misericordiae Hospital, Dublin, Ireland

Jonathan Chick
Consultant Psychiatrist, Royal Edinburgh Hospital; Senior Lecturer, Department of Psychiatry, University of Edinburgh, UK

John Cobb
Consultant Psychiatrist, The Priory Hospital, Roehampton, London, UK

Philip J. Cowen
MRC Clinical Scientist, Medical Research Council Clinical Pharmacology Unit, Littlemore Hospital, Oxford, UK

John Cutting
Consultant Psychiatrist, The Bethlem Royal Hospital, Beckenham, Kent, UK

Christopher P. Freeman
Consultant Psychotherapist, Senior Lecturer in Psychiatry, University of Edinburgh, Royal Edinburgh Hospital, Edinburgh, UK

Dennis Gath
Clinical Reader in Psychiatry, Department of Psychiatry, University of Oxford, UK

Keith Hawton
Consultant Psychiatrist and Clinical Lecturer, University Department of Psychiatry, and Warneford Hospital, Oxford, UK

Michiel W. Hengeveld
Psychiatrist, Head, Psychiatric Consultation–Liaison Service, Leiden University Hospital, The Netherlands

Peter E. Hodgkinson
Chartered Clinical Psychologist, Centre for Crisis Psychology, Skipton, Yorkshire, UK

Allan House
Consultant and Senior Lecturer in Liaison Psychiatry, General Infirmary at Leeds, Leeds, UK

Kay Redfield Jamison
Associate Professor of Psychiatry, Johns Hopkins University, Baltimore, Maryland, USA

Eve C. Johnstone
Professor of Psychiatry, Department of Psychiatry, Royal Edinburgh Hospital, Edinburgh, UK

R. Channi Kumar
Reader in Psychosomatic Medicine, Institute of Psychiatry, and Honorary Consultant Psychiatrist, The Bethlem and Maudsley Hospitals, London, UK

Laurence Mynors-Wallis
Wellcome Trust Training Fellow in Mental Health, Department of Psychiatry, University of Oxford, UK

J. Richard Newton
Senior Medical Research Fellow and Honorary Senior Registrar in Psychiatry, Royal Edinburgh Hospital, Edinburgh, UK

Philip Robson
Regional Consultant in Drug Dependency, Warneford Hospital, Oxford, UK

Nicholas Rose
Consultant Psychiatrist, Littlemore Hospital, Oxford, UK

Michael Sharpe
Clinical Tutor in Psychiatry, Department of Psychiatry, University of Oxford, UK

Pamela J. Taylor
Head of Medical Services, The Special Hospitals Service Authority, and Honorary Senior Lecturer in Forensic Psychiatry, Institute of Psychiatry, London, UK

Chris Thompson
Professor of Psychiatry, University of Southampton, Royal South Hampshire Hospital, Southampton, UK

1

Emotional problems in general practice: are psychological treatments better than drugs?

DENNIS GATH and LAURENCE MYNORS-WALLIS

Introduction

Among patients consulting their general practitioner (GP), about 25 per cent suffer from some form of psychiatric disturbance (Blacker and Clare 1987) and about a third of these have complaints that are entirely psychiatric (Goldberg and Blackwell 1970). About 95 per cent of these patients with psychiatric problems are treated within the general practice; only 5 per cent—mainly those with acute psychoses and other severe disorders—are referred to a psychiatrist. These proportions have remained unchanged despite a considerable increase in the number of consultant psychiatrist posts (Williams and Clare 1981).

In primary care, it is often difficult to use diagnostic categories of the kinds used in hospital practice (Anon. 1988). Most of the psychiatric morbidity consists of the so-called 'minor' emotional disorders. The common symptoms of these disorders have been identified by Goldberg and Huxley (1980), who studied 88 primary care patients diagnosed as having a mental disorder. Complaints of anxiety and worry were most frequent (82 patients), with despondency and sadness also common (71 patients). Other common symptoms were fatigue (71), sleep disturbance (50), and irritability (38). Psychological symptoms were often accompanied and masked by somatic symptoms, which might lead to misdiagnosis. Symptoms of anxiety and depression generally occur together—a finding that has been reported from other studies in general practice (Clare 1982; Grayson *et al.* 1987).

The minor emotional disorders can cause varying degrees of morbidity. When they occur in response to a life crisis, they may be so brief as not to merit formal psychiatric diagnosis and treatment (Tennant *et al.* 1981). In a London suburban practice, Goldberg and Blackwell (1970) found that two-thirds of a cohort of psychiatric patients recovered within six months. In another London practice, Mann and colleagues (1981) studied neurotic patients and found that one-third improved within six months, one-third had a variable and intermittent course, and one-third had chronic symptoms. Further evidence of the chronicity of some psychiatric disorders came from Dr John Fry's practice in Beckenham, Kent. Thus, in a group of patients given a first psychiatric diagnosis and followed over five years, it was found that 18 per cent of the men and 35 per cent of the women received a psychiatric diagnosis in each subsequent year (Cooper *et al.* 1969).

In addition to the minor emotional disorders, there is a significant number of more severe conditions—particularly major depression, which is found in 1 in 20 consulting patients (Blacker and Thomas 1989). There is a substantial overlap in the severity of depressive symptoms in patients treated in primary care and in those treated in hospital (Brown *et al.* 1985).

Compared with non-psychiatric patients, patients with psychiatric problems make considerable demands on their GP: they see their GP twice as often, have more physical illnesses, and have more time off work (Fry 1982).

To summarize, many patients with psychiatric symptoms are seen in general practice. Most of them are treated in the practice and not referred to specialists. Most of these patients improve quickly, but a significant number do not. Patients with emotional disorders make considerable demands on the GP's time. In the past, such patients have been treated mainly with medication, but recently, brief psychological treatments have been introduced and evaluated. This paper reviews the relative advantages and disadvantages of medication and psychological treatment in primary care.

Drug treatment

Psychotropic drugs, particularly those used to treat anxiety and depression, are widely prescribed in general practice (Williams 1979; Parish 1982). In a study of prescriptions in an Oxfordshire health centre, it was found that 16.8 per cent of registered patients received at least one prescription for a psychotropic drug during a 12 month period (Catalan *et al.* 1988).

Prescribing practices in primary care have been criticized on the grounds of polypharmacy, the prescribing of inadequate doses of antidepressants, and unnecessarily prolonged courses of treatment (Johnson 1974; Tyrer 1978). There is good evidence, however, that antidepressant medication in adequate dosage can be effective. Thus, Paykel *et al.* (1988) showed that amitriptyline was better than placebo for all but the mildest depressive disorders in general practice. In low dosage, however, antidepressants are not effective; for example, 75 mg of amitriptyline is no more effective than placebo (Blashki *et al.* 1971). Similarly, 75 mg of dothiepin is no more effective than placebo (Thompson and Thompson 1989). Depressed patients in general practice can tolerate full therapeutic doses of tricyclic antidepressants: of 227 patients treated with full dosages for major depression, only 10 per cent dropped out before six weeks because of intolerable side-effects (Blacker *et al.* 1988). Other pharmacological treatments can be useful; for example, antidepressants and beta-blockers can give symptomatic relief of anxiety (Peet 1985).

Many drug treatments, however, have disadvantages, particularly unpleasant side-effects and poor compliance, and drugs such as benzodiazepines can cause dependence. Some patients regard medication as irrelevant because they believe their illnesses to be caused by problems in the environment (Johnson 1981). There is a need, therefore, for psychological treatments to be developed as alternatives to medication.

Psychological treatments

In general practice, psychological treatments range from basic counselling techniques (for example listening, explaining, reassuring, and advising) to more elaborate treatments usually given by specialists, for example, cognitive therapy (Blackburn *et al.* 1981; Teasdale *et al.* 1984) and behavioural therapy (Ginsberg *et al.* 1984; Robson *et al.* 1984).

The most widely available psychological treatments are probably various forms of counselling. It is difficult to evaluate counselling, because it comprises a broad spectrum of therapy and is often delivered to patients with a wide range of problems. Reports of the efficacy of counselling vary. Some studies show little benefit from counselling. Thus, in one controlled trial it was found that counselling did not reduce either demands on the GP's time or the scale of psychotropic drug prescription (Ashurst 1982). In a group of patients with psychological symptoms of at least 6 months duration, neither brief dynamic psychotherapy from a psychiatrist nor counselling from a GP was more effective than the usual treatment from the GP (Brodaty and Andrews 1983). In another study, the addition of social work support to treatment as usual gave little benefit to depressed women, except for a subgroup with chronic depression and marital problems (Corney 1981). On the other hand, two uncontrolled studies describe patient improvement with counselling (Anderson and Hasler 1979; Waydenfield and Waydenfield 1980). Two other studies also demonstrate patient benefit from counselling; counselling by health visitors was effective for non-psychotic postnatal depression (Holden *et al.* 1989); and six 30 minute sessions of interpersonal counselling by a nurse practitioner was better than no treatment in relieving distress amongst patients with recent onset emotional problems (Klerman *et al.* 1987). Counselling is most likely to be of value if it is clearly defined and given to a homogeneous patient group.

In Oxford, several studies have focused on the treatment of emotional disorders in primary care. The overall aim of this research is to evaluate brief psychological treatments as possible alternatives to drugs. Some of the specific aims are to discover how far brief psychological treatments in primary care are effective, feasible, and acceptable to patients.

Brief psychological treatments in primary care

The Oxford research on these problems began with a study of brief counselling versus anxiolytic medication (Catalan *et al*. 1984). The main aim of this first study was to determine whether anxiolytic medication could be withheld without disadvantage to the patient. General practitioners were asked to select patients with new episodes of psychiatric disorder (with predominant anxiety), for which they would normally prescribe anxiolytic medication. The episode was defined as new if the onset had been in the past three months, and the patient had not taken psychotropic medication in that time.

The patients meeting these requirements were randomly allocated to anxiolytic medication or brief counselling. Anxiolytic medication was any benzodiazepine of the GP's choice (which was almost invariably diazepam). Brief counselling consisted of listening, explanation, reassurance, and advice. The two treatment groups were similar on all the main pre-treatment variables.

No significant differences were found between treatment groups at any stage of the study on any psychiatric or social outcome measure. On the General Health Questionnaire (GHQ), at the initial consultation the proportions of high scorers were: drug group 86 per cent; non-drug group 80 per cent. At the end of 4 weeks, the proportions were: drug group 38 per cent; non-drug group 46 per cent. At the end of the trial (28 weeks), the proportions were: drug group 28 per cent; non-drug group 26 per cent. For both treatment groups, these reductions in GHQ scores were highly significant at 4 weeks and again at 28 weeks. These reductions in psychiatric morbidity may have been due simply to the natural course of the disorder, and not the effects of treatment. If there was a treatment effect, however, brief counselling was at least as effective as medication.

In this study there was no evidence that withholding anxiolytics increased the consumption of tobacco, alcohol, or non-prescribed drugs. There was also no evidence that withholding anxiolytic medication increased demands on the GP's time. Finally, there were no significant differences between the two treatment groups in terms of the patients' self-rated satisfaction with the treatment received.

This study pointed to an important group of patients: those who remained unwell at the end of treatment and throughout the 28 week follow-up period, as shown by high scores on the GHQ (and also on the PSE, see below). This group seemed to merit special attention. Could they be identified at an early stage and given more effective treatment? At the initial presentation there were no clear pointers to this poor prognosis group. One month after presentation, however, they could be identified reliably by a total score above 11 on the Present State Examination (PSE) (Wing *et al.* 1974).

Problem-solving was chosen as a psychological treatment that seemed suitable for this poor prognosis group. It is a brief, non-specialized treatment that would be feasible in general practice. It focuses on personal and social problems that appear to be important in aetiology and are seen by the patient as relevant (Hawton and Kirk 1990). It is of proven value in deliberate self-harm and depression (Salkovskis *et al.* 1990; Nezu *et al.* 1989). In brief, problem-solving establishes a link between the patients' symptoms and their personal problems, and then provides a systematic way to deal with these problems. The stages of problem-solving are:

(a) clarification and definition of a problem;
(b) identification of achievable goals;
(c) generation of alternative solutions;
(d) choice of preferred solution;
(e) detailing of steps to achieve solution; and
(f) implementation and evaluation.

Problem-solving seeks not only to tackle current problems but also to equip the patient to deal with future problems that may arise.

In the second study, problem-solving was evaluated as a treatment for patients with emotional problems of poor prognosis. Patients were initially selected by the GP, as in study one, although the selection criteria included those with depressive symptoms as well as those with anxiety. Four weeks after the initial consultation, a research assistant interviewed each patient at home, and assessed him or her with the PSE. Patients scoring 12 or over on the PSE were then allocated randomly either to problem-solving given by a research psychiatrist or to treatment as usual from the GP.

A total of 113 patients were referred, of whom 47 had PSE total scores of 12 or above at 4 weeks and entered the trial. This number of patients was as predicted from the first trial described above. On random allocation, 21 were allocated to problem-solving and 26 to the GP's chosen treatment.

Trial patients were assessed on four occasions: on initial consultation (week 0); immediately pre-treatment (week 4); immediately after the end of treatment (week 12); and at follow-up (week 28). The two treatment groups were well matched for level of psychiatric morbidity before treatment.

At the end of treatment (week 11) both treatment groups had improved significantly in comparison with their pre-treatment scores. However, the total PSE score for the problem-solving group was significantly lower than for the control group. Similarly, at the 28 week follow up the mean total PSE score was significantly lower in the problem-solving group than in the control group (Catalan *et al.* 1991).

These results strongly suggest that problem-solving is a feasible and effective treatment for emotional disorders of poor prognosis in primary care. However, the treatment was given by a research psychiatrist. If problem-solving is to be widely applicable in general practice, it must be delivered by a member of the primary health care team. To answer this question, a third study is nearing completion. It has the same design as the second study, but the problem-solving is given by GPs trained in this method of treatment. The initial results of this study indicate that problem-solving can be effective when given by trained GPs.

A fourth study is under way to evaluate problem-solving as a treatment for depressive disorders in primary care. Three treatments are being compared: problem-solving; antidepressant medication; and drug placebo. The aim is to discover whether problem-solving is effective for depressive disorders in general practice, and whether it equips patients to deal with future as well as current problems and so avoid relapse.

Problem-solving would not be applicable for all emotional problems in primary care. It requires that patients should collaborate actively, and should be able to take responsibility for their actions. It does seem, however, to be a technique that could have widespread applicability.

Conclusion

Emotional disorders in general practice are common. Although most have a good prognosis, about a third do not, causing considerable psychological and social distress. Antidepressant medication has been shown to be of value for depressive conditions, but many patients refuse to take drugs. The role of medication in other conditions is much less certain. Psychological treatments normally used in hospital settings, for example cognitive therapy and behavioural treatments, can be given in general practice by specialists. They cannot be made available to the large number of patients who are treated by their GP without specialist intervention.

Problem-solving appears to be a promising psychological treatment for use in general practice. It has been shown to be effective for emotional problems of otherwise poor prognosis. It is feasible in general practice and is acceptable to patients. Twenty-five years ago it was noted that the 'cardinal requirement for improvement of the Mental Health Services in this country is not a large expansion and proliferation of psychiatric agencies but rather a strengthening of the family doctor in his therapeutic role' (Shepherd *et al.* 1966). The evidence so far suggests that problem-solving may be a valuable technique for meeting this requirement.

Clinical guidelines

1. When a specific psychiatric diagnosis can be made in general practice there is often a specific treatment that can be suggested e.g., major

depression will respond to amitriptyline taken in adequate dosage or cognitive therapy, if available. However, guidance is needed for the management of those cases which do not fit standard diagnostic criteria.

2. The majority of patients presenting to their GP with emotional problems do not need medication. Brief counselling by the GP within the context of a normal consultation is as effective as anxiolytic medication.

3. There is a group of patients (approximately one-third) who do not respond to simple measures. There are no reliable means to identify these patients at their initial consultation but one month later they will still be suffering a number of symptoms. Such patients benefit from structured psychological treatment.

4. Psychiatrists need to equip the primary health care team to manage emotional problems successfully without onward referral. (Brief problem-solving represents a simple and feasible treatment that can probably be employed by a non-psychiatrist.) A group of patients may remain, however, who would prefer to see a psychiatrist, e.g., patients with sexual problems.

References

Anderson, S. A. and Hasler, J. C. (1979). Counselling in general practice. *Journal of the Royal College of General Practitioners*, **29**, 352–6.

Anon. (1988). Classification of mental disorder in primary care. *Lancet*, **ii**, 1002–3.

Ashurst, P. (1982). Counselling in general practice. In *Psychiatry and general practice* (ed. A. Clare and M. Lader), pp. 77–88. Academic Press, London.

Blackburn, I. M., Bishop, S., Glenn, A. I. M., Whalley, L. J., and Christie, J. E. (1981). The efficacy of cognitive therapy in depression. *British Journal of Psychiatry*, **139**, 181–9.

Blacker, C. V. R. and Clare, A. (1987). Depressive disorder in primary care. *British Journal of Psychiatry*, **150**, 737–52.

Blacker, C. V. R. and Thomas, J. M. (1989). Treatment of psychiatric disorder in primary care settings. *General Hospital Psychiatry*, **11**, 216–21.

Blacker, C. V. R., Shanks, N. J., Chapman, N., and Davey, A. (1988). The drug treatment of depression in general practice. *Psychopharmacology*, **95** (Suppl.), 18–24.

Blashki, T. G., Mowbray, R., and Davies, B. (1971). Controlled trial of amitriptyline in general practice. *British Medical Journal*, **263**, 133–8.

Brodaty, H. and Andrews, G. (1983). Brief psychotherapy in family practice. *British Journal of Psychiatry*, **143**, 11–19.

Brown, G. W., Craig, T. K. J., and Harris, T. O. (1985). Depression: distress or disease? *British Journal of Psychiatry*, **147**, 612–22.

Catalan, J., Gath, D. H., Edmonds, G., and Ennis, J. (1984). The effects of non-prescribing of anxiolytics in general practice. *British Journal of Psychiatry*, **144**, 593–602.

Catalan, J., Gath, D. H., Bond, A., Edmonds, G., Martin, P., and Ennis, J. (1988). General practice patients on long term psychotropic drugs. *British Journal of Psychiatry*, **152**, 399–405.

Catalan, J., Gath, D. H., Bond, A., Day, A., and Hall, L. (1991). Evaluation of a brief psychological treatment for emotional disorders in primary care. *Psychological Medicine*, **21**, 1013–18.

Clare, A. W. (1982). Problems in psychiatric classification in general practice. In *Psychiatry and general practice* (ed. A. W. Clare and M. Lader), pp. 15–25. Academic Press, London.

Cooper, B., Fry, J., and Kalton, G. (1969). A longitudinal study of psychiatric morbidity in a general practice population. *British Journal of Preventative and Social Medicine*, **23**, 210–17.

Corney, R. H. (1981). Social work effectiveness in the management of depressed women. *Psychological Medicine*, **11**, 417–23.

Fry, J. (1982). Psychiatric illness in general practice. In *Psychiatry and general practice* (ed. A. Clare and M. Lader), pp. 43–5. Academic Press, London.

Ginsberg, G., Marks, I., and Waters, H. (1984). Cost-benefit analysis of a controlled trial of nurse therapy for neuroses in primary care. *Psychological Medicine*, **14**, 683–90.

Goldberg, D. P. and Blackwell, B. (1970). Psychiatric illness in general practice. A detailed study using a new method of case identification. *British Medical Journal*, **261**, 439–43.

Goldberg, D. P. and Huxley, P. (1980). *Mental illness in the community*. Tavistock, London.

Grayson, D. A., Bridges, K., Duncan-Jones, P., and Goldberg, D. P. (1987). The relationship between symptoms and diagnoses of minor psychiatric disorder in general practice. *Psychological Medicine*, **17**, 933–42.

Hawton, K. and Kirk, J. (1989). Problem-solving. In *Cognitive behaviour therapy for psychiatric problems: A practical guide* (ed. K. Hawton, P. Salkovskis, J. Kirk, and D. Clark), pp. 406–26. Oxford University Press, Oxford.

Holden, J. M., Sagovsky, R., and Cox, J. L. (1989). Counselling in a general practice setting; controlled study of health visitor intervention in treatment of post-natal depression. *British Medical Journal*, **298**, 223–6.

Johnson, D. A. W. (1974). A study of the use of antidepressant medication in general practice. *British Journal of Psychiatry*, **125**, 186–192.

Johnson, D. A. W. (1981). Depression: treatment compliance in general practice. *Acta Psychiatrica Scandanavia*, **63**, 447–56.

Klerman, G. L., Budman, S., Berwick, D., Weissman, M. W., Darnico-White, J., and Feldstein, M. (1987). Efficacy of a brief psychosocial intervention for symptoms of stress and distress among patients in primary care. *Medical Care*, **25**, 1078–88.

Mann, A. N., Jenkins, R., and Belsey, E. (1981). The twelve month outcome of patients with neurotic illness in general practice. *Psychological Medicine*, **11**, 535–50.

Nezu, A., Nezu, C. M., and Perri, M. G. (1989). *Problem-solving therapy for depression*. John Wiley, New York.

Parish, P. A. (1982). The use of psychotropic drugs in general practice. In *Psychiatry and general practice* (ed. A. Clare and M. Lader), pp. 65–72. Academic Press, London.

Paykel, E. S., Hollyman, J. A., Freeling, P., and Sedgewick, P. (1988). Predictors of therapeutic benefit from amitriptyline in mild depression: a general practice placebo controlled study. *Journal of Affective Disorders*, **14**, 83–95.

Peet, M. (1985). The treatment of anxiety with beta-blocking drugs. *Postgraduate Medical Journal*, **64**, (Suppl. 2), 45–9.

Robson, M. H., France, R., and Bland, M. (1984). Clinical psychologist in primary care: controlled clinical and economic evaluation. *British Medical Journal*, **288**, 1805–8.

Salkovskis, P. M., Atha, C., and Storer, D. (1990). Cognitive–behavioural problem-solving in the treatment of patients who repeatedly attempt suicide. *British Journal of Psychiatry*, **157**, 871–6.

Shepherd, M., Cooper, B., Brown, A. C., and Kalton, G. (1966). *Psychiatric illness in general practice*. Oxford University Press.

Teasdale, J. D., Fennell, M. J. V., Hibbert, G. A., and Amies, P. L. (1984). Cognitive therapy for major depressive disorder in primary care. *British Journal of Psychiatry*, **144**, 400–6.

Tennant, C., Bebbington, P., and Hurry, J. (1981). The short-term outcome of neurotic disorders in the community. *British Journal of Psychiatry*, **139**, 213–20.

Thompson, C. and Thompson, C. M. (1989). Prescribing of antidepressants in general practice. 2. A placebo controlled trial of low-dose dothiepin. *Human Psychopharmacology*, **4**, 191–201.

Tyrer, P. (1978). Drug treatment of psychiatric patients in general practice. *British Medical Journal*, **277**, 1008–10.

Waydenfield, D. and Waydenfield, S. W. (1980). Counselling in general practice. *Journal of the Royal College of General Practitioners*, **30**, 671–7.

Williams, P. (1979). The extent of psychotropic drug prescription. In *Psychosocial disorders in general practice* (ed. P. Williams and A. Clare), pp. 151–60. Academic Press, London.

Williams, P. and Clare, A. (1981). Changing patterns of psychiatric care. *British Medical Journal*, **282**, 375.

Wing, J. K., Hooper, J. E., and Sartorius, N. (1974). *The measurement and classification of psychiatric symptoms*. Cambridge University Press.

2

Psychiatrists working in general practice: which approaches are best?

NICHOLAS ROSE

Introduction

At least one in five psychiatrists does some work in general practice (Strathdee and Williams 1984). This usually involves seeing patients in an out-patient clinic shifted from the hospital. Less usually, it involves formal psychiatric meetings with the general practice team. Although these working methods go back nearly 30 years (Shepherd *et al.* 1966), their popularity has greatly increased in the past decade with local schemes starting up as a result of grass-roots interest rather than central planning. But do patients benefit? And if so, is there any particular model of working with general practitioners (GPs) that benefits patients most?

The two most important elements of working in general practice are seeing of patients, and meeting with the practice staff. This review will focus on each of these elements in turn, and look at the benefits and disadvantages for patients and professionals when compared to a traditional, hospital-centred service.

Seeing patients in a general practice setting

We do not know what proportion of patients have psychiatric consultations in a general practice setting as opposed to a hospital out-patient clinic. We do know, however, that in some places this is the norm (Tyrer 1984), and the trend is likely to increase with the development of community mental health teams. But who benefits from this? Is it really the patient, or is it, in fact, the professional?

Psychiatric clinics in general practice: good for patients?

There are two ways patients may benefit from psychiatric consultations in general practice or similar local settings. Help might be more acceptable, and it might be more effective.

Acceptability An indication of acceptability is whether patients turn up for appointments. The argument goes that the local surgery does not have a mental hospital stigma, so patients are more likely to attend.

After seeing all referrals in a general practice setting, Tyrer (1984) claimed that 19 per cent would not have attended a hospital clinic. Browning and colleagues (1987) found a 40 per cent non-attendance rate at a hospital clinic compared to 20 per cent at a local surgery. This finding is typical of a number of studies (Brown *et al.* 1988; Strathdee and Williams 1986; McLean 1986). Browning *et al.* (1987), however, found little difference in likelihood of attendance. It seems, therefore, that most studies although not all, conclude that patients are more likely to attend a local clinic compared with a hospital one.

We do not, however, know which patients would attend a local surgery but not a hospital. Are they, for example, the least or the most disabled? This is important because, clearly, if local clinics merely ensure a greater number of people with mild or transient disorders receive help, this reduces their justification. Another indication of acceptability is to ask patients where they prefer to be seen. Local general practice settings are preferred by most people. The most important reasons given are convenience, familiarity, and lack of stigma (Tyrer 1984).

Effectiveness Can patient outcome be influenced through the provision of psychiatric help in local surgeries? If we set aside the fact that patients are more likely to attend and are therefore accessible to specialist help, what other outcome measures might be improved?

First, there is evidence that for some disorders at least, earlier intervention may be possible. For example, in his study of phobic disorders (mainly agrophobia) in general practice, Marks (1985) found that the average duration of symptoms before treatment was 7 years compared with 11 years in a hospital clinic sample. There is, however, no comparable data for other disorders. Secondly, some workers have concluded that as a result of seeing patients in general practice and the associated closer contact with GPs, hospital admissions can be significantly reduced (Tyrer 1984; Hansen 1987). Why this occurred is unclear, but it is likely to have been due to a combination of factors including reductions in inappropriate admissions, availability of better community resources, and, perhaps, by earlier intervention in some cases.

Can patients lose out because psychiatrists work in general practice? If a psychiatric consultation is more available and acceptable, will referral thresholds drop in such a way as to result in minor problems swamping resources meant for major ones? Patients with severe and chronic disorders might then lose out, resources going instead to those who shout loudest.

Most studies do not, however, support this possibility. Where GP-based clinics have largely replaced hospital clinics, two studies have indicated that the proportion of severe disorders seen (predominantly schizophrenia and manic-depression) does not fall (Tyrer 1984; Browning *et al*. 1987). Where hospital and local surgery clinics coexist, the findings are less clear. One prospective study apparently showed that those with severe disorders are as likely to be seen at the general practice clinic as at the hospital clinic (Brown *et al*. 1988; Strathdee *et al*. 1990). Another retrospective case register study (Low and Pullen 1988) showed that local health centre clinics see proportionally more neurotic (36 per cent) and fewer psychotic (10 per cent) patients than their psychiatric hospital clinics (correspondingly, 22 and 25 per cent, $P<0.01$). This study, covering all new out-patient referrals in Edinburgh over a five year period, also found that local health centre clinics dealt with a larger proportion of patients receiving the label 'adjustment reaction' and 'no psychiatric diagnosis' compared with hospital clinics, amounting to 25 per cent of the health clinic and 10 per cent of psychiatric hospital clinic attenders ($P<0.01$).

Some caution must be expressed about these figures, however, because the psychiatric hospital clinics saw six times the number of patients seen in primary care, and it is unclear whether some sort of informal selection process was going on. Nevertheless, these Edinburgh findings seem in tune with many people's impressions that improved access to psychiatric help appears to result in increased referrals for less severe problems, particularly anxiety and mild depression; problems which previously, for good or ill, were contained by the general practitioner.

The apparent differences between studies in the severity of conditions seen in general practice clinics may well reflect the malleability of referral patterns. These can be influenced by what treatments are known to be locally available, by the nature of the working relationship between the GP and psychiatric team, and by local psychiatric priorities.

In summary, the main benefit for patients is that they generally prefer to be seen in local surgeries, are rather more likely to turn up, and with at least some chronic conditions, earlier intervention may be facilitated. We do not really know if hospital admissions can be reduced as a result, and we do not know whether treatment given in a general practice setting is any different in effectiveness to that given in a hospital. It is possible that psychiatric clinics in general practice may attract the referral of less severe psychiatric problems, but this has not been found in all studies and may depend on local factors.

Seeing patients in a general practice setting: good for professionals?

Psychiatrists appear to be voting with their feet if 20 per cent have chosen to spend some of their time working in a general practice setting, about two-thirds working in a shifted out-patient clinic (Strathdee and Williams 1984). What are the pay-offs for the professional staff involved?

For psychiatrists and other members of the psychiatric team, patients are more likely to turn up and the opportunity of talking with the patient's GP exists. However, this is usually at the expense of more time being spent in the car, and the inevitable administrative problems of a dispersed service. More time away from hospital may also take doctors from wards, and it may be less easy to arrange clinics in a way that facilitates the training of junior doctors.

For GPs, a psychiatric clinic down the corridor is usually welcome, even though the initiative typically comes from the psychiatric team not the GP (Strathdee 1987). What is most valued, according to one survey, is the availability of psychiatric assessment where subsequent treatment would be given either by the GP him or herself, or briefly by the psychiatrist (Strathdee 1987).

A hidden danger with general practice-based clinics is, of course, the risk that the GP's traditional role as carer for the vast majority of psychiatric problems that never get referred to a specialist (Goldberg and Huxley 1980) gets shifted on to the now more accessible psychiatric team. It is likely, however, that any effects will vary from practice to practice. Some GPs will always want to manage their own psychiatric patients, and may indeed be enabled to do this more with readily available psychiatric opinion; others may continue to be only too glad to refer patients on.

Meeting the general practice team

The second important element of a psychiatrist working in a general practice setting is the potential for meeting the primary care team. This can be informal and opportunistic, for example in the corridor or the coffee area; or formal, for example at a GP/psychiatric team liaison meeting. Strikingly, the arrangement of regular liaison meetings is relatively uncommon. In Strathdee's survey of psychiatrists spending time in general practice, only 5 per cent worked in this way (Strathdee and Williams 1984). By contrast, most researchers emphasize the importance of improved contact between GPs and members of psychiatric teams, although in most instances this is on an informal basis only (Tyrer 1984; Brown *et al.* 1988). Given that some increased informal contact between GPs and psychiatric teams will almost inevitably occur if they are working in close proximity, are formal liaison meetings of any added benefit, either for patients or for professionals?

Liaison meetings: good for patients?

The content of GP/psychiatric liaison meetings is likely to be dominated by the following: management of patients solely under GP care; management of patients under joint GP/psychiatric care; and clinical topics, mainly related to patient management (Rose 1988).

We do not know, however, whether such meetings can alter GP management behaviour in such a way as to improve psychiatric outcome for patients. All we can say is that liaison meetings provide an opportunity to get the best use out of general practice and psychiatric resources, which may in the end benefit patients in the following ways:

1. Improved GP knowledge and confidence in treating common psychiatric disorders which would not normally be referred for specialist help.
2. Improved appropriateness of referrals to specialist psychiatric services.
3. Improved co-ordination between GP and specialist where cases are shared.
4. Possibility of improved detection of hidden and severe psychiatric problems in the community with a view to offering help.

In addition, there is a 'systems' argument to consider. Most psychiatric care is provided by two health systems: general practice and specialist psychiatry. As mentioned previously, 95 per cent of patients with psychiatric disturbances remain under their GP, only 5 per cent being referred to a specialist (Goldberg and Huxley 1980). General practitioner/psychiatric liaison meetings may represent the only local forum for these systems to talk to each other regularly. In a rapidly changing health world, such links may help identify unmet needs, clarify who does what, and help identify any casualties arising from changes in health care organization.

Liaison meetings: good for professionals?

There are a number of ways in which professionals may benefit from GP/psychiatric liaison meetings. The most important relate to education, support, and effective use of resources.

Education Any regular dialogue between different professional groups dealing with similar sorts of problems is likely to be mutually educative. Members of the psychiatric and general practice teams, through meeting and discussing problems regularly, quickly come to understand each others' strengths and limitations. Information about mental health matters can be shared; for example, psychopharmacological advice or risk factors in certain disorders. Ways of conceptualizing problems along psychosocial, psychodynamic, or ethical lines can be fostered. And finally, skill techniques can be discussed or demonstrated; for example, those used in the handling of difficult consultations, or in the provision of psychological therapy.

Support Something GP and psychiatric teams share is a responsibility for problems which may be acutely worrying or alternatively chronically insoluble. Often the patients concerned are known to both teams. Liaison meetings provide an opportunity to share feelings, perhaps of doubt or helplessness. This is likely to help professionals stay with problems, and continue to contain them even if they cannot solve them. Telephone calls or any other informal contacts may also help in this. The process does not happen overnight, and to a large extent depends on trust developing between teams. It is difficult to measure the effectiveness of staff support and supervision activities. However, feedback from GPs does suggest that the supportive function of liaison meetings is of particular value (Rose 1988).

Effective use of resources Demand for psychiatric resources is always likely to outstrip supply. Thus, any arrangement which helps to rationalize who gets what is worth consideration. Liaison meetings can help this process in three main ways: as a means of sharing information about available psychiatric resources; as a forum to discuss appropriateness of referral, thus giving some control over the front door into the psychiatric service; and as a setting in which to decide on shared care or GP care arrangements in a way that makes the best use of the combined resources of the psychiatric and general practice teams.

Can liaison meetings be problematic for professionals? Five potential difficulties stand out. First, closer face-to-face contact between GP and psychiatric teams could worsen relationships. Usually, however, they can only get better. One study of 10 practices over a year found no occurrences of relationships getting worse (Rose 1988). Second, liaison meetings may only preach to the converted. A minority of GPs invariably miss meetings and it is possible that it is they whom the psychiatric team would most like to influence. Third, meetings are expensive in professional time which might more usefully be used seeing patients. Fourth, shared care arrangements, or arrangements where the patient is discussed with but not seen by the psychiatric team may blur the question of who is ultimately responsible for the patient's clinical care. This issue needs to be clearly discussed and clarified between the GP and psychiatric teams. Fifth, there is the risk that in helping GPs to manage psychiatric problems more effectively, psychiatrists risk ending up only dealing with patients who have psychotic disorders.

In summary, regular meetings between general practice and psychiatric teams can be problematic, but have many potential benefits for psychiatrists and professionals. However, these have yet to be linked with improved patient outcome measures and so it is hard to judge their cost effectiveness. Evaluative studies need to be done to assess the effectiveness of different models of working, and in particular the nature of information and skills that can most usefully be passed on to GPs. One important question, for example, is whether diagnostic and treatment skills can be altered through the influence of regular liaison meetings. In addition, more work needs to be done in order to evaluate the different contributions to liaison meetings by different disciplines, in particular, nursing, psychology, and social work. Until all this has been carried out, any decision about setting up liaison meetings must be based on clinical and, perhaps more importantly, management judgement as to whether it is thought that this arrangement will enable a mental health service to be delivered more efficiently and effectively. It could be argued that with the movement toward a community care service and the development of community psychiatric teams, working relationships with general practice have never been more important.

The clinical experience of one psychiatrist

What follows is a brief description of a combined liaison meeting/general practice clinic service developed in 10 practices in South Buckinghamshre. This service complemented a hospital-based out-patient clinic, and resulted from a questionnaire sent to GPs which identified a need for closer contact between GPs and psychiatrists.

The liaison meetings were monthly and usually included doctors, nurses, and a social worker. The purpose of the meetings was twofold: (1) to help the GP team manage psychiatric problems themselves; and (2) to enable appropriate use to be made of a community psychiatric team. Meetings involved psychiatric topics or case discussion, including the effect patients have on individual staff or teams as a whole. In addition to liaison meetings, new referrals were seen in the surgery.

Liaison meetings

Two of the liaison meetings stopped within six months, one because of a major internal crisis in the practice, entirely unrelated to the visits of the psychiatric team; the other because of resistance from an influential senior partner who was never persuaded of the need for meetings. The remaining eight practices were continuing to meet regularly three years later.

During the first year of operation the content of liaison meetings was recorded, with approximate timings for different activities (these are summarized in Table 1).

As can be seen, gathering information about patients and subsequent discussion on the understanding of problems and the management options available took about half of the time. The rest of the time was equally divided between the provision of information and the discussion of skill-related, ethical or psychodynamic issues influencing either the general practice team as a whole, or one individual in particular.

Table 1   Content of general practice liaison meetings and proportion of time spent on different activities

Information collection Mainly about patients	23%
Discussion: patient factors Understanding psychosocial, physical, and psychodynamic factors. Management including appropriateness of psychiatric referral.	26%
Discussion: general practice team factors Doctor/patient relationship (e.g., coping with demanding patients). Using psychological techniques/skills (e.g., simple counselling). Within-practice dynamics (e.g., dumping on partners). Ethical issues (e.g., confidentiality).	25%
Educational information Prescribing. Hidden psychiatric disorders. Assessing risk. Local psychiatric resources, etc.	26%

Out-patient clinic

The first hundred patients seen as new referrals in the clinics in general practice were compared with the first hundred patients discussed in the liaison meeting but not seen because the GP wished to retain full care (see Table 2).

Patients with neurotic depression were greatly over-represented in the assessed group, mainly because of GP concerns about the associated suicide risk. By contrast, patients with personality disorders tended not to get referred, although advice on their management was often sought. Most sexual problems, substance abuse patients, and those thought to probably have an adjustment reaction were either managed by the GP, or after discussion were referred to other agencies. Patients with functional psychoses were about equally assessed and discussed. Where they were not seen by the psychiatric team, this was usually because of patient reluctance, which in many cases changed over time, leading to eventual assessment.

The chronicity of the disorders as measured by duration of at least one year was similar in both groups, 68 per cent for those discussed, 61 per cent for those assessed. The history of past psychiatric consultations was also similar, correspondingly 42 per cent and 34 per cent, a figure similar to that found in other liaison attachment schemes (McKechnie *et al.* 1981; Creed and Marks 1989).

Table 2   Comparison of patients' problems
discussed in general practice liaison meetings with
those of patients seen in a general practice clinic
(n = 100 in each group)

	Discussed %	Assessed %
Functional psychosis	12	10
Organic state	2	2
Neurotic disorder	20	60
Personality problem	18	5
Eating disorder	7	6
Sexual problem	6	0
Substance abuse	8	3
Adjustment reaction	13	5
No psychiatric diagnosis	6	6
Prevention advice	8	3

Clinical guidelines

Closer working arrangements with GPs are likely to be helped by psychiatric
teams being sectorized according to general practices. This allows the
development of long-term relationships and commitments. Given this
linkage, certain choices present themselves.

1. Which model of working?

Should liaison meetings be combined with a shifted out-patient clinic?
Should this clinic replace or complement the hospital clinic? Choice is likely
to be limited by local conditions. It is more cost effective to have liaison
meetings and clinics in large health centres. So if the psychiatric team covers
many practices, or practices which are small, many patients may still have
to be seen in hospital clinics, although more local alternatives, such as community hospitals, can often be found. In all practices, space is almost
always at a premium, and one may have to be flexible by seeing patients
when GPs are not holding surgeries.

Shifted out-patient clinics pose a number of procedural problems which
have to be overcome. For example, the mechanism of referral and booking,
and the raising and storing of psychiatric files. In contrast, liaison meetings
depend more on sympathetic attitudes than procedures or physical space.
They require persuasiveness and persistence to establish but are logistically
simple to run.

In the end, the psychiatrist has to offer arrangements that take into account local conditions, his or her own capacity to deliver a particular service, the attitude of local GPs, and what he or she believes will benefit patients most.

2. Liaison meetings: choices

Once the decision has been made to set up regular liaison meetings the following points will need to be addressed:

When and how often? Monthly meetings lasting about one hour are recommended. Meetings every two months lack continuity, particularly if a meeting is missed, and weekly or fortnightly meetings are often impractical, especially if there is a network of other liaison meetings. Most meetings are over lunchtime for convenience, but 8.30–9.30 a.m. with surgeries starting after this are worth considering.

Who should go? It is important that the meetings involve experienced psychiatrists, either at senior registrar or consultant level. This helps both continuity and credibility and facilitates the supervisory and educational components of meetings.

Psychologists and community psychiatric nurses also increasingly work in general practice settings (Marks 1985; CPNA 1990), either as members of a multidisciplinary psychiatric team or less commonly as independent practitioners directly accountable to the GP. In either case, they should be included in liaison meetings, as should a psychiatric social worker and community occupational therapist or rehabilitation worker, if available. In practice, not all of these disciplines need go to every liaison meeting and three representatives are usually sufficient.

In addition, health visitors, district nurses, and practice counsellors can be invited from the general practice team. From time to time, others can also be invited, for example, those responsible for running services for the elderly, for children, or for substance abusers. In this way, liaison meetings can be a vehicle for more general contact between primary care and the whole range of mental health services, including those in the voluntary sector.

What should be the content of meetings? The meetings will need some form of structure, and also a co-ordinator, either from the psychiatric or general practice team. The overall aim of meetings should be agreed by all, and a regular review of their value carried out at least once a year. Each meeting should have an agenda, either agreed in advance or at the beginning of the meeting. Generally speaking, liaison meetings develop a style of their own which reflects what people want to get out of them, and it is for this reason that they should not be too rigid. Needs change over time, and what may have started as a need for information may go on to fulfil quite different requirements.

Are new skills required to conduct liaison meetings?   Important skills needed include those for group facilitation, for teaching, and for supervision. But perhaps most importantly, the psychiatrist needs to be credible to primary care staff in his or her role as offering a shop window for psychiatry, able not only to persuade them of the value of a psychological approach, but also to get them to use it in their work.

3.   Psychiatric out-patient clinics in general practice: choices

Once it has been decided that it is practical to start a clinic in general practice, some practical issues will need to be addressed.

How will the clinic operate?   New bookings can go through the psychiatric team, which gives them some control, or can be made directly by GPs. In either case, a referral letter is highly desirable because it focuses the minds of all concerned. The psychiatrist may want to build in the option of joint appointments, seeing some patients with their GP. Most psychiatrists seeing new patients in general practice will open a psychiatric file, but for briefer consultations a note in the GP file may be an alternative. Follow-up appointments may be limited because of space. Finally, procedures will need to take into account the different funding arrangements for patients in budget-holding and non-budget holding practices, whilst making sure that patients get the same quality of service in each.

Who should go?   Psychiatric clinics in general practice can be staffed by a psychiatrist, or by a multidisciplinary team under the clinical responsibility of a psychiatrist. The latter arrangement has been shown to be acceptable to GPs (Rose 1988), particularly if the multidisciplinary team also attends general practice meetings.

How often?   It has been shown in one study that non-attendances for first and subsequent appointments were over twice as common in general practice/psychiatric clinics held monthly compared to those held weekly (Strathdee *et al.* 1990). Monthly clinics also give limited flexibility in responding to more urgent referrals and in follow-up. Ideally then, clinics should be weekly or at least fortnightly. However, in many instances this may not be practical because of distance, numbers of practices covered, or rates of referral.

References

Brown, R. M. A., Strathdee, G., Christie-Brown, J. R. W., and Robinson, P. H. (1988). A Comparison of referrals to primary care and hospital out patient clinics. *British Journal of Psychiatry*, **153**, 168–73.

Browning, S. M., Ford, M. F., Goddard, C. A., and Brown, A. C. (1987). A psychiatric clinic in general practice; a description and comparison with an out patient clinic. *Bulletin of the Royal College of Psychiatrists*, **11**, 114–17.

CPNA (Community Psychiatric Nursing Association) (1990). *The 1990 national survey*, pp. 10–13. CPNA Publications, CPNA, 25 Hall Park Avenue, Liversedge, West Yorkshire WS15 7EH, UK.

Creed, F. and Marks, B. (1989). Liaison psychiatry in general practice: a comparison of the liaison-attachment scheme and shifted out patient clinic models. *Journal of the Royal College of General Practitioners*, **39**, 514–17.

Goldberg, D. and Huxley, P. (1980). *Mental illness in the community: the pathway to psychiatric care*. Tavistock, London.

Hansen, V. (1987). Psychiatric services within primary care. *Acta Psychiatrica Scandinavia*, **76**, 121–8.

Low, L. B. and Pullen, I. (1988). Psychiatric clinics in different settings: a case register study. *British Journal of Psychiatry*, **153**, 243–5.

Marks, I. (1985). Controlled trial of psychiatric nurse therapists in primary care. *British Medical Journal*, **280**, 1181–4.

McLean, E. M. (1986). Patterns of collaboration: general discussion. In *Mental illness in primary care settings* (ed. M. Shepherd, G. Wilkinson, and P. Williams). Tavistock, London.

McKechnie, A. A., Philip, A. E., and Ramage, J. G. (1981). Psychiatric services in primary care: specialised or not? *Journal of the Royal College of General Practitioners*, **31**, 611–14.

Rose, N. (1988). An analysis of the first 100 psychiatric–general practice liaison meetings. Poster presentation. Quarterly meeting, Royal College of Psychiatrists.

Shepherd, M., Cooper, B., Brown, A. C., and Kalton, G. W. (1966). *Psychiatric illness in general practice*. Oxford University Press.

Strathdee, G. (1987). Primary care—psychiatry interaction: a British perspective. *General Hospital Psychiatry*, **9**, 102–10.

Strathdee, G. and Williams, P. (1984). A survey of psychiatrists in primary care: the silent growth of a new service. *Journal of the Royal College of General Practitioners*, **34**, 615–18.

Strathdee, G. and Williams, P. (1986). Patterns of collaboration. In *Mental illness in primary care settings* (ed. M. Shepherd, G. Wilkinson, and P. Williams). Tavistock, London.

Strathdee, G., Brown, R. M. A., and Doig, R. J. (1990). Psychiatric clinics in primary care; the effect on general practitioner referral patterns. *Social and Psychiatric Epidemiology*, **25**, 85–100.

Tyrer, P. (1984). Psychiatric clinics in general practice. An extension of community care. *British Journal of Psychiatry*, **145**, 9–14.

3

New antidepressants: have they superseded tricyclics?

PHILIP J. COWEN

Introduction

Tricyclic antidepressants

Treatment with antidepressant drugs still offers the best hope of symptomatic relief of depressive disorders of any severity. In spite of their limitations, tricyclic antidepressants have been the mainstay of drug treatment for the last three decades. Should this still be the case?

Tricyclics are undoubtedly more effective than placebo in the treatment of depressive illness, but up to one-third of patients may fail to respond, even with an adequate dose and duration of treatment. In addition, tricyclics have a slow onset of therapeutic activity and many undesirable side-effects (Table 3), which may make treatment unpleasant or even untenable in patients who are striving to carry on their usual activities.

Perhaps the most serious adverse effect of tricyclics is their toxicity in deliberate overdose (Cassidy and Henry 1987), which is always a significant hazard in depressed patients. Although careful assessment and restriction of the quantity of drug prescribed can minimize this risk, it seems inevitable that some patients will attempt to take their lives in this way. In addition, accidental overdose (most likely, of course, in children) can have similarly disastrous consequences.

New antidepressants

It will be clear from the above that the treatment of depression with tricyclics is far from satisfactory. Over the last 10 years several new antidepressants have been introduced which are claimed to have advantages over tricyclics. Two of these claims can be substantiated: (1) many of the new antidepressants lack significant anticholinergic activity; and (2) the new antidepressants are far less toxic in overdose. Before more detailed discussion of some of the newer drugs is undertaken, two general points will be made concerning efficacy and longer-term toxicity and use in prophylaxis.

Efficacy Establishing the efficacy of antidepressant drugs is not easy, because depression tends to remit spontaneously and there is a significant

Table 3 Some adverse effects of tricyclic antidepressants

Pharmacological action	Adverse effect
Muscarinic receptor blockade (anticholinergic)	Dry mouth, tachycardia, blurred vision, glaucoma, constipation, urinary retention, sexual dysfunction, cognitive impairment
α_1-Adrenoceptor blockade	Drowsiness, postural hypotension, sexual dysfunction, cognitive impairment
Histamine H_1-receptor blockade	Drowsiness, weight gain
5-$HT_{2/1C}$-receptor blockade	Weight gain
Membrane stabilizing properties	Cardiotoxicity, seizures

response to placebo. For these reasons it is generally considered that controlled clinical trials of new antidepressant drugs should include comparison with placebo. Even with the long-established tricyclics, however, the data are far from clear-cut. Morris and Beck (1974), for example, reviewed 93 controlled trials in which tricyclic antidepressants had been compared with placebo but in only 61 of these studies was the tricyclic therapeutically superior.

For most of the new drugs there is reasonable evidence of superior antidepressant efficacy compared to placebo. However, it is also necessary to compare new antidepressant compounds with established drugs and design of these studies is problematic. None of the newer antidepressants is claimed to be superior to established compounds; their antidepressant efficacy is usually regarded as being equivalent. However, to detect small (but clinically significant differences) in efficacy between two active drugs, very large numbers of patients are needed and these are not often achieved.

It is also important to note that the controlled trials discussed above may well involve patients who are only moderately depressed. For example, the subjects studied are often out-patients (sometimes recruited by newspaper advertisements) who lack clinical features, such as pronounced suicidal ideation or severe melancholic symptoms, which would make a double-blind trial of an untried medication unethical. There is, therefore, concern that the efficacy of new drugs has not been established as equivalent to tricyclics in the more seriously ill patients (Danish University Antidepressant Group 1990).

Long-term toxicity Although it is clear that most of the newer antidepressant drugs are safer in overdose than tricyclic antidepressants, the longer-term toxicity of the newer drugs is less well established. The reason for this is that rare (but hazardous) side-effects may only become apparent after considerable patient exposure. Such adverse reactions have led to the withdrawal of two of the more recently introduced antidepressants, nomifensine and zimelidine.

Use in prophylaxis   It is now increasingly recognized that many depressed patients require long-term antidepressant treatment to prevent recurrent depressive episodes. There is some evidence, albeit rather limited, that tricyclics are effective for this indication and some 5-hydroxytryptamine (5-HT) uptake blockers appear to be as well (Montgomery *et al.* 1988). For other new antidepressants, however, strong evidence of longer-term efficacy has not yet accumulated.

Table 4   Cost of some new antidepressant drugs compared with amitriptyline (the doses are roughly equivalent)

Drug	Dose (mg)	Approx. cost of 30 days treatment
Amitriptyline	150	£2.16
Mianserin	60	£9.88
Amoxapine	150	£13.50
Lofepramine	210	£16.00
Trazodone	300	£20.10
Fluvoxamine	100	£25.00
Sertraline	50	£26.51
Fluoxetine	20	£27.44
Paroxetine	20	£33.90

Extended treatment with antidepressants has significant cost implications and most of the newer antidepressants are much more expensive than standard tricyclics (Table 4). Even in purely economic terms, however, the prevention of depressive illness would seem to be a cost effective exercise and the major issue therefore is whether the better tolerability and safety of newer antidepressants gives them a signficant advantage over standard tricyclics in prophylactic regimes. Such studies have yet to be carried out. Below, some of the pharmacological and clinical properties of the newer antidepressants are considered in more detail. Some new drugs are not considered because they were not licensed for use in the UK at the time this article was prepared.

5-HT uptake blockers

The term '5-HT uptake blocker' denotes a class of compounds whose acute pharmacological activity is confined to the blockade of neuronal uptake of 5-hydroxytryptamine (5-HT) (Table 5). Although most conventional tricyclic antidepressants inhibit 5-HT uptake they also inhibit the uptake of noradrenaline as well as antagonizing a variety of other neurotransmitter receptors. This latter property is largely responsible for their wide range of adverse effects (Rudorfer and Potter 1989) (Table 3).

Therapeutic efficacy

There is extensive evidence from controlled clinical trials that fluoxetine and fluvoxamine are superior to placebo in the treatment of major depression. The same is probably true of sertraline and paroxetine, although clinical experience with these latter compounds is less extensive (Boyer and Feighner 1991*a*). In general, selective 5-HT uptake inhibitors seem to be as effective as amitriptyline or imipramine, although in some studies the reference 'tricyclic' was superior to the 5-HT uptake inhibitor and in others the reverse was the case (Boyer and Feighner 1991*a*).

The onset of therapeutic effect of 5-HT uptake inhibitors may be slower than that seen with tricyclics; a meta-analysis showed that significant improvement over placebo was seen at 4 weeks with tricyclics but at 6 weeks with 5-HT uptake inhibitors (Bech 1990). In addition, as mentioned above, there is relatively little data available on the therapeutic efficacy of selective 5-HT uptake inhibitors in severely ill patients and in these subjects it remains possible that the drugs are less effective than tricyclics (Danish University Antidepressant Group 1990).

For those patients who respond to 5-HT uptake blockers, there is fairly consistent evidence that the clinical response to 5-HT uptake inhibitors can be maintained during longer-term maintenance and prophylactic treatment. Comparisons with tricyclics for this indication are few in number and have yielded equivocal findings (Boyer and Feighner 1991*a*).

Tolerance

Generally, but not invariably, controlled studies have shown that selective 5-HT uptake inhibitors are better tolerated than tricyclics such as amitriptyline and imipramine. In particular, 5-HT uptake blockers lack anticholinergic properties and are less cardiotoxic than tricyclics (Boyer and Feighner 1991*b*). In addition, they do not produce cognitive impairment (Hindmarch 1990). They do, however, have a characteristic side-effect profile which includes nausea and other gastrointestinal disturbances, together with insomnia and anxiety, which can be particularly troubling in the initial stages of treatment (Schatzberg *et al.* 1987; Boyer and Feighner

Table 5 Pharmacological actions and adverse effects of some newer antidepressants

Antidepressant	Anticholinergic activity	Sedation	Toxicity in overdose	Other side-effects
5-HT uptake inhibitors				
Fluoxetine	0	0	<TCA	Nausea, vomiting, anxiety, insomnia,
Fluvoxamine	0	0	<TCA	headache, reduced appetite, sweating,
Sertraline	0	0	<TCA	skin rash, generalized allergic reaction
Paroxetine	0	0	<TCA	(rare), seizures (rare).
Modified tricyclics				
Amoxapine	++	0	=TCA	Extrapyramidal movement disorders, endocrine disturbances, seizures (usually in overdose).
Lofepramine	+	0	<TCA	Anxiety, insomnia, seizures (rare).
Sedating antidepressants				
Mianserin	0	++	<TCA	Cognitive impairment, postural hypotension, bone marrow. depression (rare), seizures (rare).
Trazodone	0	++	<TCA	Cognitive impairment, postural hypotension, priapism (rare).

TCA = tricyclic antidepressant; 0 = not present; + = modest; + + = marked.

1991*b*). In the longer term, however, anxiety symptoms may respond well. Appetite reduction (in the absence of nausea) may also occur.

There may be some clinical differences between the 5-HT uptake inhibitors, with fluoxetine, for example, more likely to provoke stimulation and insomnia, whereas fluvoxamine has been more associated with somnolence and nausea (Boyer and Feighner 1991*b*). However, these apparent differences may reflect the fact that equivalent dosing regimes for the different compounds have not yet been clearly established.

Toxicity

There seems little doubt that 5-HT uptake blockers are considerably safer than tricyclics in overdose (Cassidy and Henry 1987). During clinical treatment few serious adverse reactions have been reported and when these have occurred (for example, seizures or allergic skin rash), the incidence does not seem any higher than that seen with tricyclics (Boyer and Feighner 1991*b*). There are reports, however, of serious drug interactions with agents that also potentiate 5-HT neurotransmission such as monoamine oxidase inhibitors (MAOIs), L-tryptophan, and lithium (Boyer and Feighner 1991*c*). Co-administration of these compounds with 5-HT uptake blockers can result in a syndrome of 5-HT neurotoxicity, with confusion, myoclonus, seizures, and hyperpyrexia. With appropriate caution it is possible to co-administer lithium and 5-HT uptake inhibitors, but combination with MAOIs is contra-indicated (Boyer and Feighner 1991*c*). Fluoxetine poses particular difficulties in this respect because its active metabolite, nora-fluoxetine, has a 7–15 day half-life (Schatzberg *et al.* 1987). Accordingly, it is recommended that 5 weeks must elapse between stopping fluoxetine treatment and starting an MAOI.

Suicidal ideation

A report by Teicher *et al.* (1990) linked fluoxetine treatment of depressed patients with the development of violent suicidal preoccupations. Whether or not this disturbing phenomenon is attributable to fluoxetine is still controversial. It is possible that in some predisposed patients the stimulant and arousing effects of fluoxetine could be translated into greater suicidal thinking and activity. Consistent with this is a report of the emergence of suicidal ideation in depressed patients receiving treatment with desipramine (Damluji and Ferguson 1988), a relatively selective noradrenaline (NA) uptake inhibitor, which also has some activating properties. In addition, a recent meta-analysis of 17 double-blind clinical trials demonstrated no significant increase in suicidal thoughts or acts in patients receiving fluoxetine compared to those on tricyclic antidepressants or placebo (Beasley *et al.* 1991).

Modified tricyclic drugs

Two more recently introduced tricyclic antidepressants, amoxapine and lofepramine, have been claimed to have significant advantages over conventional tricyclic drugs while retaining the therapeutic efficacy of the established compounds (Table 5).

Amoxapine

Pharmacology and efficacy   Amoxapine is a derivative of the neuroleptic drug loxapine. Its pharmacological effects include inhibition of noradrenaline uptake and antagonist effects at post-synaptic dopamine receptors. Its affinity for other neurotransmitter receptors appears somewhat less than conventional tricyclics (Rudorfer and Potter 1989). Controlled trials demonstrate that the efficacy of amoxapine is equivalent to that of standard tricyclic antidepressants and its onset of action may be earlier (Bernstein 1986), perhaps due to its dopamine receptor blocking properties.

Implications of dopamine receptor blockade   The dopamine antagonist properties of amoxapine may also be of value in treating depressive psychosis where controlled studies have shown that the combination of a tricyclic and neuroleptic is more effective than either drug administered alone (Anton *et al.* 1986). However, the use of fixed-dose combination treatment limits prescribing flexibility. In addition the dopamine receptor blocking properties of amoxapine have been associated with extrapyramidal movement disorders, including tardive dyskinesia (Rudorfer and Potter 1989).

Tolerance and toxicity   The tolerance of amoxapine may be a little better than conventional tricyclics, with less anticholinergic effects described in clinical trials. In overdose, however, amoxapine has a high toxicity, with a substantial risk of seizures and a mortality rate of the same order as older tricyclics (Rudorfer and Potter 1989).

Lofepramine

Pharmacology and efficacy   Lofepramine is a modified tricyclic antidepressant which is a selective inhibitor of noradrenaline uptake, but has less affinity than standard tricyclics for monoamine and muscarinic receptors (Lancaster and Gonzalez 1989). Placebo-controlled studies of lofepramine are few in number but suggest that the drug has antidepressant activity in depressed patients. In addition, several controlled comparisons have demonstrated that the antidepressant efficacy of lofepramine appears to be equivalent to that of amitriptyline and imipramine (Lancaster and Gonzalez 1989).

Tolerance and toxicity   Clinical experience has shown that lofepramine appears to be much less toxic in overdose than conventional tricyclics, which clearly gives it a major advantage in certain groups of patients (Cassidy and Henry 1987). In addition, lofepramine is generally better tolerated than standard tricyclics and produces fewer anticholinergic effects, with little evidence of cognitive impairment (Hindmarch 1990). Clinically, most patients experience lofepramine as somewhat activating. This can cause problems in the early stages of treatment when increased anxiety and sleep disturbance may be apparent. Following longer-term treatment, however, as with 5-HT uptake blockers, associated anxiety symptoms may respond well.

Sedating antidepressants

The newer antidepressants described have greater pharmacological selectivity than conventional tricyclic antidepressants and, in particular, have much less antagonist effect at α_1-adrenoceptor and histamine H_1-receptor sites. Probably because of this they lack sedating properties and indeed may be experienced by some patients as activating. This can cause difficulties in depressed subjects who present with severe sleep disturbance and associated anxiety because these symptoms may not improve or can even get worse in the early stages of treatment. There is, therefore, a place in the treatment of depression for antidepressant drugs with sedative properties and two newer drugs, mianserin and trazodone, will be considered under this heading (Table 5).

Mianserin

Pharmacology and efficacy   Mianserin is a quadracyclic compound that does not inhibit the uptake of noradrenaline or 5-HT but instead antagonizes a variety of monoamine receptors. Its antidepressant activity has been attributed to blockade of presynaptic α_2-adrenoceptors which would be expected to facilitate the release of noradrenaline. It has a pronounced sedative action presumably due to blockade of α_1-adrenoceptors and antihistaminic effects (Barnes and Bridges 1982). The efficacy of mianserin is not easy to judge because of the diverse nature of the trials in which it has been tested and the wide dosage range employed (30–120 mg) (Barnes and Bridges 1982; Rudorfer and Potter 1989). A reasonable overall conclusion is that mianserin is more effective than placebo and, in moderately depressed patients at least, is as effective as conventional tricyclic treatment (Barnes and Bridges 1982; Rudorfer and Potter 1989).

Tolerance and toxicity   Mianserin lacks significant anticholinergic activity and is not cardiotoxic. It is much safer in overdose than conventional tricyclics (Cassidy and Henry 1987). The α_1-adrenoceptor antagonist properties of mianserin can result in a modest degree of postural hypotension together with over-sedation and cognitive impairment (Barnes and Bridges 1982; Hindmarch 1990). Excessive weight gain is not infrequently encountered.

Perhaps the major concern in the use of mianserin, however, is its association with rare but occasionally fatal cases of agranulocytosis (Chaplin 1986). The risk of this reaction is disputed, with figures from different countries varying markedly. In the UK, for example, it has been claimed that the rate may not significantly exceed that seen with tricyclics (Inman 1988). The risk appears to be greatest in elderly patients during the first few weeks of mianserin treatment and routine blood count monitoring is recommended during this period (Committee on Safety of Medicines 1989).

Trazodone

Pharmacology and efficacy Trazodone is a triazalopyridine derivative which has complex pharmacological actions including blockade of both α_1-adrenoceptors and 5-HT-receptors (Brogden *et al.* 1981). It also possesses some 5-HT uptake blocking properties but it is doubtful if these are manifest at usual clinical doses (Fuller and Wong 1987). *In vivo*, trazodone is metabolized to *m*-chlorophenylipiperazine, which is a 5-HT$_1$-receptor agonist (Fuller *et al.* 1981). There is extensive evidence from controlled clinical trials that trazodone is more effective than placebo in the treatment of depression and studies in moderately depressed patients suggest that its antidepressant efficacy is equal to that of conventional tricyclics (Brogden *et al.* 1981; Barnes and Bridges 1982). Some concern has been expressed, however, about its effectiveness in more seriously ill patients (Rudorfer and Potter 1989).

Tolerance and toxicity Trazodone has prominent sedating effects and dosage needs to be titrated cautiously early in treatment (Schatzberg *et al.* 1987). Not surprisingly, such sedation may be linked to impaired psychological performance (Hindmarch 1990). Trazodone does not produce anticholinergic side-effects but its α_1-adrenoceptor antagonist properties may be associated with a moderate degree of postural hypotension (Barnes and Bridges 1982; Rudorfer and Potter 1989). A rare but potentially serious autonomic side effect is priapism, which has sometimes required pharmacological or even surgical intervention (Schatzberg *et al.* 1987; Rudorfer and Potter 1989). Trazodone is less cardiotoxic than conventional tricyclics but disturbances in cardiac rhythm have sometimes been reported during treatment (Rudorfer and Potter 1989). In overdose, trazodone lacks the toxicity seen with standard tricyclics and appears to lower seizure threshold somewhat less (Potter and Rudorfer 1989).

Clinical guidelines

1. Newer antidepressants have significant advantages over conventional tricyclics in terms of improved tolerance, less impairment of psychological performance, and increased safety in overdose. However, their longer-term toxicity has yet to be fully evaluated and their efficacy in severely ill patients is not as well established as that of tricyclics. Newer antidepressants are much more expensive than standard tricyclics.

2. Depressed patients who have previously shown a good clinical response and acceptable tolerance with tricyclic antidepressant treatment should continue to receive these drugs both for the management of new episodes of depression and for longer-term prophylaxis.

3. In other patients with moderate depressive disorder and a good short term prognosis, initial drug treatment should probably be with a newer antidepressant, either a 5-HT uptake inhibitor or lofepramine. If anxiety and sleep disturbance are likely to pose problems in the initial stages of treatment, consideration can be given to the short-term co-administration of a benzodiazepine, or to the use of a more sedating compound such as mianserin or trazodone.

4. Patients with more severe depressive disorders, particularly those being managed as in-patients should probably receive a standard tricyclic antidepressant unless side-effects make this impracticable or if the risk of suicide through overdose cannot be minimized.

References

Anton, R. F., Hitri, A., Diamond, B. I., and Shelhorse, M. (1986). Amoxapine treatment of psychotic depression: dose effect and dopamine blockade. *Journal of Clinical Psychiatry*, **47**, 32–6.

Barnes, T. R. E. and Bridges, P. K. (1982). New generation of antidepressants. In *Drugs in psychiatric practice* (ed. P. J. Tyrer), pp. 219–48. Butterworth, London.

Beasley, Jr. C. M., *et al.* (1991). Fluoxetine and suicide: a meta-analysis of controlled trials of treatment for depression. *British Medical Journal*, **303**, 685–91.

Bech, P. (1990). A meta-analysis of the antidepressant properties of serotonin reuptake inhibitors. *International Review of Psychiatry*, **2**, 207–11.

Bernstein, J. G. (1986). Amoxapine: rapid onset and clinical use. *Journal of Clinical Psychiatry*, **47**, 3–8.

Boyer, W. F. and Feighner, J. P. (1991*a*). The efficacy of selective serotonin re-uptake inhibitors in depression. In *Selective serotonin reuptake inhibitors* (ed. J. P. Feighner and W. F. Boyer), pp. 89–108. Wiley, Chichester.

Boyer, W. F. and Feighner, J. P. (1991*b*). Side effects of the selective serotonin re-uptake inhibitors. In *Selective serotonin reuptake inhibitors* (ed. J. P. Feighner and W. F. Boyer), pp. 133–52. Wiley, Chichester.

Boyer, W. F. and Feighner, J. P. (1991c). Pharmacokinetics and drug interactions. In *Selective serotonin reuptake inhibitors* (ed. J. P. Feighner and W. F. Boyer), pp. 81–8. Wiley, Chichester.

Brogden, R. N., Heel, R. C., Speight, T. M., and Avery, G. S. (1981). Trazodone: a review of its pharmacological properties and therapeutic use in anxiety and depression. *Drugs*, **21**, 406–76.

Cassidy, S. L. and Henry, J. (1987). Fatal toxicity of antidepressant drugs in overdose. *British Medical Journal*, **295**, 1021–4.

Chaplin, S. (1986). Bone marrow depression due to mianserin, phenylbutazone, oxyphenbutazone, and chloroamphenicol: Part II. *Adverse Drug Reactions and Acute Poisoning Reviews*, **3**, 181–96.

Committee on Safety of Medicines (1989). Mianserin and white blood cell disorder in the elderly. *Current Problems*, No. 25.

Damluji, N. and Ferguson, J. M. (1988). Paradoxical worsening of depressive symptomatology caused by antidepressants. *Journal of Clinical Psychopharmacology*, **8**, 347–9.

Danish University Antidepressant Group (1990). Paroxetine: a selective serotonin reuptake inhibitor showing better tolerance, but weaker antidepressant effect than clomipramine in a controlled multicenter study. *Journal of Affective Disorders*, **18**, 289–99.

Fuller, R. W. and Wong, D. T. (1987). Serotonin reuptake blockers in vivo and in vitro. *Journal of Clinical Psychopharmacology*, **7**, 36S–43S.

Fuller, R. W., Snoddy, H. D., Mason, N. R., and Owens, J. E. (1981). Disposition and pharmacological effects of m-chlorophenylpiperazine in rats. *Neuropharmacology*, **20**, 155–62.

Hindmarch, I. (1990). Antidepressants: the implications of the cognitive and psychomotor effects in the elderly. *International Clinical Psychopharmacology*, **5**(Suppl. 3), 57–60.

Inman, W. H. (1988). Blood disorders and suicide in patients taking mianserin or amitriptyline. *Lancet*, **333**, 90–2.

Lancaster, S. G. and Gonzalez, J. P. (1990). Lofepramine: a review of its pharmacodynamic and pharmacokinetic properties and therapeutic efficacy in depressive illness. *Drugs*, **37**, 123–40.

Morris, J. B. and Beck, A. T. (1974). The efficacy of antidepressant drugs: a review of research (1958–1972). *Archives of General Psychiatry*, **30**, 667–74.

Montgomery, S. A., *et al.* (1988). The prophylactic efficacy of fluoxetine in unipolar depression. *British Journal of Psychiatry*, **153**(Suppl. 3), 69–76.

Rudorfer, M. V. and Potter, W. Z. (1989). Antidepressants: a comparative review of the clinical pharmacology and therapeutic use of the 'newer' versus the 'older' drugs. *Drugs*, **37**, 713–38.

Schatzberg, A. F., Dessain, E., O'Neil, P., Katz, D. L., and Cole, J. O. (1987). Recent studies on selective serotonergic antidepressants: trazodone, fluoxetine and fluvoxamine. *Journal of Clinical Psychopharmacology*, **7**, 44S–49S.

Teicher, M., Glod, C., and Cole, J. O. (1990). Emergence of intense suicidal preoccupation during fluoxetine treatment. *American Journal of Psychiatry*, **147**, 207–10.

4

Manic-depressive illness: what role does psychotherapy have in management?*

KAY REDFIELD JAMISON

Introduction

Manic-depressive, or bipolar illness, encompasses a wide continuum of mood disorders and temperaments. These range in severity from cyclothymia—characterized by pronounced but not totally debilitating changes in mood, behaviour, thinking, sleep, and energy levels—to extremely severe, life-threatening, and psychotic forms of the disease. Manic-depressive illness is both strongly genetic and recurrent; left untreated, individuals with this disease can expect to experience many, and generally worsening, episodes of depression and mania during their lifetimes. Importantly, however, most individuals who have manic-depressive illness are normal most of the time; that is, they maintain their reason and their ability to function personally and professionally. Often lethal, prior to the availability and widespread use of lithium, one person in four or five with manic-depressive illness committed suicide (Goodwin and Jamison 1990). It is a relatively common illness; approximately 1 per cent of the population will suffer from the more severe forms of it during their lifetimes; at least that many again will experience milder variants, such as cyclothymia (Depue et al. 1981; Weissman et al. 1988).

Highly effective treatments exist for manic-depressive illness. Lithium has radically altered the course and consequences of the disease, allowing most patients to live reasonably normal lives. In recent years, anticonvulsant medications, such as carbamazepine and sodium valproate, have provided important alternative treatments for patients unable to take, or unresponsive to lithium. Although biological factors predominate in the aetiology of manic-depression and pharmacological ones in its treatment, nonetheless, the primary manifestations of manic-depressive illness are behavioural and psychological, with profound changes in perception,

---

* Based on chapters written by K. R. Jamison in F. K. Goodwin and K. R. Jamison (1990). *Manic depressive illness*. Oxford University Press, New York.

attitudes, personality, mood, and cognition. Psychological interventions, therefore, can be of unique value to patients undergoing such devastating changes in the way they perceive themselves and the world, and are perceived by others.

Given the efficacy of lithium, clinicians may minimize the value of psychotherapy and their own role in the treatment of manic-depressive illness. Vasile and colleagues (1987), for example, found that psychiatrists and mental health professionals discouraged psychodynamic psychotherapy in the treatment of affectively ill patients. Patients themselves, by contrast, often find psychotherapy a potent adjunct to lithium. In the one study in which patients were actually asked, twice as many patients as therapists thought psychotherapy was helpful to them in remaining compliant with medication (Jamison *et al.* 1979). There are also times when clinicians encounter manic-depressive patients who are not on medication, and psychotherapy may have to serve as the sole treatment (e.g., patients who refuse medication or women who stop lithium during their pregnancies). In addition, psychotherapy in conjunction with lithium, may be the treatment of choice for breakthrough depressions in patients prone to antidepressant-induced cycling. There are also theoretical grounds to expect that psychotherapy may help to ameliorate some of the stress-related precipitants of manic and depressive episodes. Such intervention may, one hopes, temper the progression of the natural course of the illness (Post *et al.* 1986).

Psychological support for the treatment of manic-depressive patients ranges from a few minutes with the prescribing physician to combined use of individual and group psychotherapy. Usually, a general psychiatrist or psychopharmacologist treats patients on lithium, typically within a limited time period of 20 to 30 minutes, every several weeks. Although comprehensive psychotherapeutic work cannot take place in such a context, the doctor can create an emotionally supportive atmosphere, be aware of, and focus on the general psychological issues involved in taking lithium and having an affective illness, and encourage patients to express their concerns. Providing a therapeutic relationship of this kind increases the likelihood of lithium compliance and makes it more probable that the patient will be referred for formal psychotherapy when there is a need for it. Formal, structured psychotherapy—for example, cognitive or interpersonal therapy —best follows control of acute episodes.

Review of the literature

Several modalities of psychotherapy are available to the clinician wishing to work with patients on lithium. Informal psychological treatments include the supportive role of the physician in medication management, educational models that convey medical information about manic-depressive illness and lithium through lectures, handouts, films, or information-giving groups, and self-help groups run by patients for themselves. A wide range of theoretical orientations governs more formal psychological treatments, such as individual, group, family, and joint psychotherapy or some combination of these.

Little is known about the appropriateness of any of these modalities or orientations for a particular patient or type of problem. For example, we do not know if cognitive behavioural therapy is the most effective approach for lithium compliance problems or if psychodynamic therapy is the most appropriate for the interpersonal sequelae of manic-depressive illness. We also do not know if group therapy is more useful for problems of illness denial and self-esteem or if individual therapy is more appropriate for short-term crisis intervention in problems, such as suicidal behaviour, or for the long-term treatment of idiosyncratic, intrapsychic problems. Several studies involving couple and family therapy have been done, and most are summarized in Table 6.

The available clinical reports on combined lithium treatment and psychotherapy are shown in Table 6. Few of the studies used comparison groups for the analysis of treatment outcomes or pre-test and post-test measures. The substantial methodological shortcomings in most of these investigations make meaningful interpretation virtually impossible but the clinical observations made by the therapists conducting the studies (noted in Table 6) are conceptually very useful.

Better controlled studies (Cochran 1984; Glick *et al.* 1985) indicate that both individual and family psychotherapy, when combined with lithium, seem to result in a better clinical outcome and increased lithium compliance. Jacobs (1982) also described specific cognitive therapy techniques to help patients overcome post-manic and post-depressive dysphoria. Wulsin and colleagues (1988) reviewed the literature on group therapy in manic-depressive illness and argue for its efficacy, stressing that several aspects of the illness make group therapy a likely intervention: (1) the patient's need for information about the illness; (2) the patient's need for lithium and other medical management; and (3) the interpersonal difficulties that develop during the course of manic-depressive illness.

Issues for psychotherapy in manic-depressive patients

For many patients, lithium is an uncertain treatment imposed on an uncertain illness, a problematic treatment for a problematic disease. For many, life before lithium can be likened to a kite on a string in exceedingly unpredictable winds. Lithium gives some control over the winds, but often it is not complete. And therein lies much of the disappointment and frustration. Clinicians frequently define successful outcome very differently than patients do. The clinician looks at certain types of evidence—fewer or no

Table 6 Clinical reports on combined lithium treatment and psychotherapy

Study	Therapy/Design/Patients (*n*)	Results	Clinical observations
Fitzgerald (1972)	Family therapy and lithium, eclectic: educational, emphasis on communications. Clinical study: no comparison, 25 BP (bipolar patients), index hospitalization for mania.	No systematic follow-up.	Family therapy can help manic patients: (1) take lithium; (2) prevent relapses; and (3) improve verbal communication within the family (i.e., replace the role of mania in the expressions of anger and frustration).
Davenport *et al.* (1977)	Group I: psychodynamic couples' group + lithium, 4 times wk; 12 BP.	No re-hospitalizations; no marital failures.	At follow-up, Group I patients significantly improved on social functioning and family interactions *vs.* Group III; better on family interaction *vs.* Group II; no significant differences between Groups II and III. Strong recommendation for co-therapist model. Marital dynamics: (1) fear by both spouses of recurrence; (2) sense of helplessness; (3) need to control all affect and defend against closeness; (4) use of massive denial; and (5) themes related to early parental loss and failure to grieve.
	Group II: NIMH out-patient dept. Lithium maintenance 4 times mth; crisis treatment as needed, 11 BP.	2 re-hospitalized; 5 marital failures.	
	Group III: Referral to community clinic or private care. *Note:* Marriages intact at time of discharge from index hospitalization for all; follow-up period variable for all (2–10 yr post-hospitalization), 42 BP.	16 re-hospitalized; 10 marital failures; 3 suicides.	

Shakir et al. (1979)	Group therapy + lithium; interpersonal, interactional. Clinical study; compared group means of patients' pre- and post-hospitalization records. 75 min 4 times wk; lithium dispensed during last 30 min of group. 15 BP — 13 M, 2 F; mean age: 43. Range: 19–63. Mth on lithium prior to group: 21.	Pre-group: 16 wk/yr in hospital. Post-group: 3 wk/yr in hospital. 2 yr follow-up	Group themes: (1) initial scepticism about lithium and group therapy; (2) complaints about loss of well-being due to lithium; (3) denial of problems; (4) projection of responsibility for lithium deficiency on to psychiatrists; (5) concerns: recurrence, illness chronicity, social adjustment, and social acceptance.
Rosen (1980)	Group therapy + lithium; directive. Clinical study; no comparison, pre- or post-test; 90 min 4 times wk for 4 wk. 25 mixed patients, 12 M, 13 F, mean age: 36. Range: 25–27.	No systematic follow-up; over 2.5 yr 8 patients remained in group.	Group themes: (1) lithium; (2) concerns about manic-depressive illness; (3) lowered morale when member hospitalized. Described lively atmosphere where 'therapist works not so much to stimulate interaction as to moderate it'.
Volkmar et al. (1981)	Group therapy + lithium; interpersonal, interactional. (Continuation of group therapy outlined in Shakir et al. 1979.)		Group themes as reported in Shakir et al. (1979). Also, cessation of lithium use from denial of illness, lack of information, lack of support. 'It is not clear whether the high rate of compliance is secondary to the effects of group therapy per se or to the close follow-up the patients received or to the interaction between the two.'

Table 6 (cont.)

Study	Therapy/Design/Patients (*n*)	Results	Clinical observations
Cochran (1984)	Individual therapy + lithium; 6 sessions cognitive–behavioural therapy vs. standard clinic care; 26 BP, 13 in each condition.	Significantly more lithium compliance and fewer hospitalizations in psychotherapy group.	Although two groups did not differ significantly in number of affective episodes, psychotherapy patients were less likely to be hospitalized.
Glick *et al.* (1985)	In-patient family intervention. Compared with a standard multi-model inpatient treatment; 12 BP families in therapy group, 8 in comparison.	At 18 mth follow-up, significantly fewer re-hospitalized and better work/social role functioning in family therapy group.	Clinical outcome was best. Family intervention was the most effective when the patient's family was assessed as low on 'patient rejection' and patient was compliant with medication.
Kripke & Robinson (1985)	Group out-patient, problem-solving. Blood levels and prescriptions monitored. VA setting. Started with 13 M, 1 F; 8 remained at 12th yr of group. 90 min every 2 wks.	Decreased hospitalization rates, improved socio-economic functioning.	Leader's attempts at a psychodynamic and introspective focus worked less well than problem-solving.

Haas *et al.* (1988); Spencer *et al.* (1988)	In-patient family intervention, psychoeducational, 169 psychiatric patients (50 affective disorder) randomly assigned to family intervention or none (with an increase in individual psycho-therapy for the latter). Average of six 45–60 min sessions.	At discharge, improved hospital treatment outcome for the family intervention group, especially in females with affective disorders. At 6 and 18 mth follow-up, a better outcome was also seen among the schizophrenics in this group.	Attitudes of families toward treatment improved in the study group. Later follow-ups showed greater openness to social support, less patient rejection and family burden.
Miklowitz *et al.* (1988); Goldstein & Miklowitz (unpublished)	Behavioural family treatment (including education, communication training, problem-solving) + lithium. 8 BPs (bipolar patients) treated for 9 mth with naturalistic follow-up. Compared with 23 BP families receiving lithium only.	13% of behavioural + lithium group relapsed. 70% of lithium-alone group relapsed.	

From Goodwin and Jamison 1990.

hospitalizations or little or no need for adjunctive neuroleptics and anti-depressants—and finds lithium effective. The patient who continues to experience disruptive and upsetting mood swings is likely to interpret the same evidence in much more equivocal terms. In essence, physicians more often focus on the successes of lithium, that is, the contrasts with untreated illness. Patients, although living with the successes, live with the failures and disappointments as well. Patients also see the contrast, but they find themselves comparing the dramatic improvements with day-to-day discontents. The improvements tend to be forgotten, and with time, the seriousness of the illness is denied. The day-to-day discontents then emerge as the compelling factor in feelings about lithium.

Therapeutic issues of general concern involve many areas of patients' adjustment to having manic-depressive illness, including fears of recurrence, denial of the illness, discrimination of normal from pathological moods, adjustment to the effects of the illness on normal developmental tasks, and concerns about the hereditability of the disease.

Fear of recurrence

The worst fear for most manic-depressive patients is recurrence of the illness. Many patients maintain a deep and fatalistic pessimism, entwined with denial and optimism, about again becoming manic or depressed. Some patients become preoccupied with such fears of recurrence and are almost illness-phobic. They become unduly self-protective and hyper-alert for signs of an impending episode. These concerns are often reflected in the process of learning to differentiate normal from abnormal moods and states. A sense of having a decreasing tolerance for affective episodes is a concern that is usually secondary to the stress of the illness and to the large amount of psychological energy consumed by earlier bouts. Patients, often with good cause, fear that their families and friends will grow increasingly intolerant with each new recurrence. Manic-depressive illness also takes a severe toll on other relationships, professional activities, and the individual's ability to handle the emotional stress of the affective episodes.

Discrimination of abnormal moods

Problems in learning to discriminate normal from abnormal moods are common throughout the psychotherapy of bipolar patients. Because of the intensity of their emotional responses, many manic-depressive patients fear that a normal depressive reaction will deepen into a major episode and that a state of well-being will escalate into hypomania or mania. Many common emotions range across several mood states, spanning euthymia, depression, and hypomania. For example, irritability and anger can be a part of normal human existence or, alternatively, can be symptoms of both depression and hypomania. Tiredness, sadness, and lethargy can be due to normal circumstances, medical causes, or clinical depression. Feeling good, being productive and enthusiastic, and working hard can be either normal or pathognomonic of hypomania. These overlapping emotions can be confusing and arouse anxiety in many patients, who may then question their own judgement and become unduly concerned about recurrences of their affective illness.

Helping the patient discriminate normal from abnormal affect is common in psychotherapy. The patient must learn to live within a narrower range of emotions yet master the skill of using those emotions with greater subtlety and discretion. Closely related to the discrimination of moods is the slow, steady process involved in patients learning to unravel what is normal personality from what the illness has superimposed upon it—turbulence, impulsiveness, lack of predictability, and depression.

Use of mood charts

Mood charting by patients can provide invaluable information about seasonal and other patterns of moods, psychological and biological correlates of mood swings, and responsiveness to treatment, including possible worsening of the illness due to treatment (e.g., increased cycling induced by antidepressant therapy). Addressing this last point, Post and colleagues (1986, p. 198) stressed the utility of identifying 'critical psychosocial stresses and areas of sensitivity in a given patient, which appear to be temporally related to repeated episodes of affective illness'. The administration of the Visual Analogue Scale is straightforward, requiring little time on the part of the patient. The patient is given sheets of paper, each with a 100 mm line, anchored by a 1 ('worst I've ever felt') on the left (or the bottom) and 100 ('best I've ever felt') on the right (or the top). The patient is then asked to put a mark across the line at the point most representative of his or her overall mood (or whatever other variable, e.g., energy or anxiety level, is being assessed) for the day. To control for diurnal variations in mood and behaviour, ratings should be done at approximately the same time of day or evening. Significant life events and additional medications required should be noted on the rating sheet. After completion, the dated form should be placed aside to avoid contamination from earlier ratings. A 100 mm ruler is used to score the patient's mood ratings. The results can then be graphed, with time plotted along the horizontal axis and mood ratings, from 1 to 100, plotted along the vertical axis. In some instances, patients can do their own graphing (see Chapter 18).

Graphing these mood ratings is useful not only in noticing patterns of mood and treatment response but also in giving patients a sense of control, instilling a feeling of collaborative effort, and underscoring the importance of systematic observation. It also provides a relatively objective basis for persuading patients when their treatment regimens require modification. At the beginning of treatment, other patients' charts can be used to illustrate different patterns of mood fluctuation and the importance of daily mood ratings in diagnostic and treatment decisions. One essential teaching point is that there is often an uneven, sawtooth nature to the recovery pattern. Predicting occasional serious relapses on the way to remission is important in minimizing serious, potentially lethal discouragement during a high-risk (i.e., transitional) period.

Education

Patients often express resentment at how little information they receive about manic-depressive illness and its treatment. Most affective disorders clinics and some practitioners routinely provide formal and informal education to patients and families through lectures, books, articles, pamphlets, discussion groups, videotapes, and ongoing communication between clinicians and patients. However, this is the ideal scenario and not the prevailing one. Clinicians, in whatever setting, have an obligation to engage patients in a continuing process of education and informed consent. Patients vary considerably in their ability to assimilate information about their medications and illness, and they need to participate actively in the treatment process. Too often, a physician becomes a unilateral advocate of lithium and other maintenance medications, and this frequently leads to an adversarial rather than collaborative effort. Patients should be encouraged to question their clinicians about diagnosis and treatment, to discuss their concerns about undue delays in getting the desired results, and to seek second opinions where appropriate. If the treating physician or psycho-therapist disparages second opinions or consultations, patients should be encouraged to challenge this opinion and obtain the consultation anyway.

Patient education and informed consent, integrally bound, require an informed clinician. The chronic and highly recurrent nature of manic-depressive illness should be emphasized and re-emphasized to the patient. Charts can be used to illustrate the high relapse rate and worsening course in the untreated illness (for example, Squillace *et al.* 1984), as well as the dramatic effect of lithium on the course of manic-depressive illness (Baastrup and Schou 1967). Patients should be encouraged to read about the illness and its treatment, as well as being given very specific information about medications and potential risks and side-effects.

As with any form of treatment, safety and efficacy and the risk of no treatment should be outlined. Special attention should be paid to the discussion of potential dangers of antidepressant use in manic-depressive patients (such as induction of mania and worsening of the natural course of the illness, i.e., shortening of the cycle length).

Prodromal symptoms

Patients need to be alerted to the symptoms of impending episodes. Changes in sleep patterns are particularly important, because they precede mania, and sleep loss may precipitate it. Environmental changes leading to insomnia (e.g., anxiety, excitement, grief) or others (such as hormonal changes, travel, drugs) can lead to mania through sleep deprivation. Wehr and Goodwin (1987) advise that manic-depressive patients should be warned that a single night of unexplainable sleep loss should be taken as an early warning of possible impending mania. They further suggest counselling patients to avoid situations likely to disrupt sleep and advise clinicians to consider prescribing clonazepam to prevent significant sleep loss. The regularization of circadian rhythms through the regularization of meals, exercise, and other activities should also be stressed to patients.

The collaborative nature of the patient–clinican relationship is central to effective treatment. Not only must patients be taught about the natural course and symptoms of manic-depressive illness, they should also be actively encouraged to express to their doctor concerns about their illness and treatment.

Involvement of family

Family members and close friends often find that the educational information given to patients is also useful to them. Families are, of course, in a unique position to observe the behaviour and moods of bipolar patients. Education about the illness can increase the awareness and acceptance of patients and underscore the family's role in encouraging the patient to take prescribed medications and to live sensibly. Waiting for symptom-free intervals between episodes to discuss the meaning and nature of manic-depressive illness allows for education and collaborative decision-making in a less emotionally charged atmosphere. Family members, in addition to being educated about medications and the illness, should be informed about the importance of recognizing the early signs and symptoms of manic and depressive episodes. Changes in sleep patterns, sexual and financial behaviour, mood (expansiveness or undue enthusiasm, pessimism, and hopelessness), involvement in excessive numbers of projects, and changes in judgement are all highly characteristic of impending affective episodes. Often, these changes are first noted by family members and can be crucially important to the patient in early intervention. Strategies for contacting the clinician should be determined, if possible, during times when the patient is normal. As far as is feasible, general contingency plans and agreements should be made in advance to cover possible emergencies (e.g., suicidal thinking and behaviour), hospitalization plans for mania, and financial protection for the patient and family during hypomanic and manic episodes.

The potential problems of violation of confidentiality are substantial and need to be discussed openly with patients and their families.

Medication compliance

Medication is the central treatment for manic-depressive illness, not an adjunctive one. From time to time, lithium non-compliance becomes a major theme in the therapy of many patients (Jamison *et al.* 1979; Jamison and Akiskal 1983). Confusion often arises because the illness itself, as well as its pharmacological treatments, can affect cognition, perception, mood, and behaviour. Although the emphasis in this chapter is on psychotherapeutic issues for patients treated with lithium, most of the discussion is applicable to issues that arise from patients maintained on other medications such as carbamazepine or sodium valproate. Psychotherapeutic sessions often involve concerns about being on medication in general and lithium in particular. Lithium's effectiveness in ameliorating the illness is not always welcome, because it deprives some patients of energy and much sought-after highs, and, additionally, can burden them with bothersome side-effects.

The consequences of non-compliance are clinically equivalent to those of untreated or inadequately treated manic-depressive illness: recurrence and intensification of affective episodes that are often accompanied by interpersonal chaos, alcohol and drug abuse, personal anguish and family disruption, financial crises, conjugal failure, psychiatric hospitalization, violence, and suicide. This point may be obvious, but it is frequently ignored. Unlike unresponsiveness, however, non-compliance is reversible and can be changed through experience, education, learning, and psychotherapy.

Generally, compliance appears to increase with age, which coincides with a period of increasing risk of episodes recurring. The first year after initiation of lithium treatment is a particularly high-risk period for stopping lithium against medical advice. A constellation of related mood variables also seems to predict non-compliance: missing the highs, elevated mood in its own right, and a history of grandiose delusions. There is some evidence that patients experiencing proportionately more manic than depressive episodes before starting lithium, are more likely to be non-compliant.

Guidelines for maximizing lithium compliance are summarized in Table 7. Sackett and associates (1985) emphasize the 'uniformly dismal performance of clinicians in predicting and assessing compliance in their patients. They found that detecting which patients are non-compliant could be done more quickly, less expensively, and about as well, simply by asking patients about taking their medication as by using drug levels. To increase the reliability of patients' responses, they suggest an interview format that makes the

Table 7 Guidelines for maximising lithium compliance

Monitor compliance	Regular lithium levels. Enquire frequently. Encourage queries and concerns from patients and families.
Side-effects	Forewarn. Treat aggressively (especially hypothyroidism and tremor). Minimize lithium level.
Education	Early symptoms of mania and depression. Unremitting and worsening course of (untreated) manic-depressive illness.
Medication	Minimize number of daily doses. Pillboxes (7-day), especially if on two or more drugs. Involve family members in administering if appropriate. Written information about lithium and side-effects (limited), patient-titration of lithium level.
Adjunctive psychotherapy	
Self-help groups	

From Goodwin and Jamison (1990).

admission of non-compliance socially acceptable, for example, 'Most people have trouble taking all of their pills. Do you have any trouble taking all of yours?' The physician should order lithium level determinations regularly and enquire frequently about possible problems with compliance and concerns about the medication.

As emphasized earlier, psychotherapy is important in the treatment of manic-depressive illness, specifically in encouraging lithium compliance. Lithium patients tend to place a far greater value on adjunctive psychotherapy than do clinicians, and non-compliant patients have been shown to regard psychotherapy as highly useful in helping them adhere to a regimen of lithium treatment (Jamison *et al.* 1979). Consistent with these observations are findings that show that patients treated with cognitive therapy more often took their lithium as prescribed than did patients who did not receive psychotherapy (Cochran 1982). Because of their compliance, the psychotherapy patients also had fewer hospitalizations.

Clinical management of suicidal manic-depressive patients

The most reliable method of preventing suicide in manic-depressive patients is to treat the underlying illness effectively. Most manic-depressive patients are at high risk for suicide, but some are at even higher risk because of a family history of suicide or because of their clinical state. Early and accurate diagnosis is critical to the identification of these especially vulnerable patients. It is also important because the danger of suicide appears to be greatest in the initial phases of the illness, a pattern of risk that also influences treatment decisions, such as the timing of lithium maintenance therapy and the timing and frequency of supportive psychotherapy.

Explicit information about manic-depressive illness, its treatment, and suicide is particularly important when dealing with suicidal patients, who may feel profound hopelessness and be cognitively impaired. Whenever feasible, information should be provided to them in both oral and written form. One of the first messages needing clear communication concerns the limits on confidentiality between suicidal patients and their therapists. This message becomes very significant for patients who are paranoid, irritable, and hostile or are experiencing mixed states and rapidly fluctuating moods. Other explanations and predictions for the patient and, where appropriate, the family are listed in Table 8.

It is important to communicate consistently that, although manic-depressive illness is serious, it can be treated successfully in the vast majority of cases. Left untreated, however, particularly early in the illness, it often results in suicide. The clinician must explain to the patient and family that denial or recurrence is common, but it can also be dangerous. Such an explanation predicts feelings and thereby lends credence to the clinician's recommendations.

The patients must take lithium and other medications as prescribed, and be assured that the drug can often work as effectively against depression as it does against mania but that there is usually more time delay before it has an effect in the prevention or treatment of depression. The patients (and clinician) should not be discouraged by this delay nor assume that depression and suicidal feelings are inevitable.

It is important to communicate explicitly that many side-effects occurring with medication can be ameliorated; others cannot. The clinician should be specific about possible side-effects and about how transitory or permanent they are likely to be.

Patients who are on antidepressant medication should be warned that the time course for a drug response may lead to a discrepancy between what

Table 8 Communication to patients and families

General issues	Written information whenever possible. Ways of contacting clinician. Limits on confidentiality. Postpone major life decisions. Treatable nature of affective illness.
Medication issues	Many effective medications available. Imperative to take medications as prescribed. Instructions should be in writing. Side-effects usually transient and/or treatable.
When to contact physician	Worsening of suicidal ideation. Worsening of symptoms, especially: • sleep loss; • agitation, severe restlessness; • delusions; and • feelings of violence, impulsivity. Problems with medication compliance.
Alcohol and drugs	Worsen sleep and judgement. Potentiate prescribed medications. Undermine efficacy of medication. Increase likelihood of mixed states.
Recovery issues	High-risk nature of recovery period. Recovery likely to be frustrating and tumultuous. Sawtooth curve pattern. Time course and recovery pattern with antidepressants.

From Jamison (1988).

their physician sees as improvement and what they themselves are experiencing. For example, the physician and family may see improvement because the patient has more energy and is sleeping better, and because the patient's face and body are more animated. These changes generally occur before improvements in mood and thinking, changes that are likely to be more important to the patient. Predicting this discrepancy in perceptions can lessen some of the patient's discouragement, which is particularly important because at this stage in the illness the patient is at high risk for suicide.

As the patient's condition begins to improve, the clinician may find it necessary to explain that a particularly frustrating and difficult period lies ahead and that temporary setbacks are common. The clinician should inform the patient and family that recovery from a suicidal depression is exceptionally difficult and likely to be filled with ups and downs, successes and setbacks.

The patients should be aware that alcohol generally worsens depression, interferes with sleep, impairs judgement, and potentiates the effects of other medications. The patient should be advised also to avoid significant social or personal changes when depressed and to obtain a leave of absence from school or work rather than quit (Winokur *et al.* 1969).

The treatment of manic-depressive illness remains one of the true successes of modern medicine. The basics of lithium therapy are now widespread in clinical practice. Far less widespread are the use of adjunctive psychotherapy, the subtle use of medication titration, mood charting, systematic patient and family education, and referral to patient and family support groups. The development of specific approaches for psychotherapy with manic-depressive patients is essential.

Clinical guidelines

1. Manic-depressive illness is treated most effectively with a combination of lithium or other medications and adjunctive psychotherapy. Although not all patients need psychotherapy, most can benefit from one of its many forms—individual, group, or family. Participation in a self-help group can supplement or supplant psychotherapy.

2. Educating patients and their families about manic-depressive illness—its symptoms, course, and treatment—is essential. It encourages compliance, increases the sense of participation in the treatment process, and helps in the recognition of impending episodes.

3. Patients should be educated about those symptoms that are most likely to predict dangerous relapses (e.g., sleep loss, delusions, agitation, and feelings of violence); if and when these symptoms occur they should be treated aggressively.

4. Suicide is a very real possibility in manic-depressive illness; patients and their families should be aware of this and, whenever possible, contingency plans should be made in advance.

5. Several steps can be taken to increase lithium (or other medication) compliance: minimize, whenever possible, the medication level and aggressively treat the drug side-effects; track the patients' compliance, and discuss openly misgivings about treatment; educate the patients about the role of lithium and anticonvulsants in attenuating the course and consequences of manic-depressive illness.

References

Baastrup, P. C. and Schou, M. (1967). Lithium as a prophylactic agent: Its effect against recurrent depression and manic-depressive psychosis. *Archives of General Psychiatry*, **16**, 162–72.

Cochran, S. D. (1982). Strategies for preventing lithium noncompliance in bipolar affective illness. Doctoral dissertation, University of California, Los Angeles.

Cochran, S. D. (1984). Preventing medical noncompliance in the outpatient treatment of bipolar affective disorders. *Journal of Consulting and Clinical Psychology*, **52**, 873–8.

Davenport, Y. B., Ebert, M. H., Adland, M. L., and Goodwin, F. K. (1977). Couples group therapy as an adjunct to lithium maintenance of the manic patient. *American Journal of Orthopsychiatry*, **47**, 495–502.

Depue, R. A., Slater, J. F., Wolfstetter-Kausch, H., Klein, D., Goplerud, E., and Farr, D. (1981) A behavioural paradigm for identifying persons at risk for bipolar depressive disorder: A conceptual framework and five validation studies. *Journal of Abnormal Psychology Monograph*, **90**, 381–437.

Fitzgerald, R. G. (1972). Mania is a message: Treatment with family therapy and lithium carbonate. *American Journal of Psychotherapy*, **26**, 547–53.

Glick, I. D., *et al.* (1985). A controlled evaluation of inpatient family intervention: Preliminary results of the six-month follow-up. *Archives of General Psychiatry*, **42**, 882–6.

Goodwin, F. K. and Jamison, K. R. (1990). *Manic-depressive illness*. Oxford University Press.

Haas, G. L., *et al.* (1988). Inpatient family intervention: A randomized clinical trial. II. Results at hospital discharge. *Archives of General Psychiatry*, **45**, 217–24.

Jacobs, L. I. (1982). Cognitive therapy of post-manic and post-depressive dysphoria in bipolar illness. *American Journal of Psychotherapy*, **36**, 450–8.

Jamison, K. R. (1988). Suicide prevention in depressed women. *Journal of Clinical Psychiatry*, **49**, 42–5.

Jamison, K. R. and Akiskal, H. S. (1983). Medication compliance in patients with bipolar disorders. *Psychiatric Clinics of North America*, **6**, 175–92.

Jamison, K. R., Gerner, R. H., and Goodwin, F. K. (1979). Patient and physician attitudes toward lithium: Relationship to compliance. *Archives of General Psychiatry*, **36**, 866–9.

Kripke, D. F. and Robinson, D. (1985). Ten years with a lithium group. *McLean Hospital Journal*, **10**, 1–11.

Miklowitz, D. J., Goldstein, M. J., Nuechterlein, K. H., Snyder, K. S., and Mintz, J. (1988). Family factors and the course of bipolar affective disorder. *Archives of General Psychiatry*, **45**, 225–31.

Post, R. M., Rubinow, D. R., and Ballenger, J. C. (1986). Conditioning and sensitization in the longitudinal course of affective illness. *British Journal of Psychiatry*, **149**, 191–201.

Rosen, A. M. (1980). Group management of lithium prophylaxis. Abstract of a paper presented at the annual meeting of the American Psychiatric Association.

Sackett, D. L., Haynes, R. B., and Tugwell, P. (1985). *Clinical epidemiology: A basic science for clinical medicine*. Little, Brown & Co., Boston.

Shakir, S. A., Volkmar, F. R., Bacon, S., and Pfefferbaum, A. (1979). Group psychotherapy as an adjunct to lithium maintenance. *American Journal of Psychiatry*, **136**, 455–6.

Spencer, J. H. (1988). A randomized clinical trial of inpatient family intervention: iii. Effects at 6-month and 18-month follow-ups. *American Journal of Psychiatry*, **145**, 1115–21.

Squillace, K., Post, R. M., Savard, R., and Erwin-Gorman, M. (1984). Life charting of the longitudinal course of recurrent affective illness. In *Neurobiology of mood disorders* (ed. R. M. Post and J. C. Ballenger), pp. 38–59. Williams & Wilkins, Baltimore.

Vasile, R. G., *et al.* (1987). A biopsychosocial approach to treating patients with affective disorders. *American Journal of Psychiatry*, **144**, 341–4.

Volkmar, F. R., Shakir, S. A., Bacon, S., and Pfefferbaum, A. (1981). Group therapy in the management of manic-depressive illness. *American Journal of Psychotherapy*, **35**, 226–34.

Wehr, T. A. and Goodwin, F. K. (1987). Can antidepressants cause mania and worsen the course of affective illness? *American Journal of Psychiatry*, **144**, 1403–11.

Weissman, M. M. *et al.* (1988). Affective disorders in five United States communities. *Psychological Medicine*, **18**, 141–53.

Winokur, G., Clayton, P. J., and Reich, T. (1969). *Manic depressive illness*. Mosby, St Louis, Virginia.

Wulsin, L., Bachop, M., and Hoffman, D. (1988). Group therapy in manic-depressive illness. *American Journal of Psychotherapy*, **42**, 263–71.

5

Management of mood disorder in adults with brain damage: can we improve what psychiatry has to offer?

ALLAN HOUSE

Introduction

Brain damage acquired in adult life is a major health problem. For example, each year in the UK 100 000 admissions result from traumatic head injury and approximately 120 000 people suffer from a stroke. In an average health district there will be about 2000 adults who have brain damage from one of these causes; others will have similar problems arising from cerebral anoxia, the sequelae of CNS infection, and a wide range of rarer insults to this most sensitive of organs.

Of all physical illnesses, one might expect those which cause brain damage to produce most in the way of mood disturbance, and yet in many psychiatry textbooks the problem is barely mentioned. This deficit in the literature is matched by clinical ones; few psychiatrists feel confident in assessing or treating the emotional problems of patients with brain damage and even fewer would profess to special expertise in the area.

This chapter outlines the areas in which our clinical practice could be improved, and indicates the treatments which deserve consideration because they might be effective. None has been evaluated using rigorous research designs, but even so I would argue that there are reasonable grounds for recommending psychiatric involvement in the assessments and treatments suggested.

Improving our descriptive vocabulary

Brain damage poses a problem of description and definition, because it produces emotional disorders which are rarely found in other settings. I will give two examples of syndromes which are common following brain damage and yet not as clearly delineated as they should be.

Emotionalism

All psychiatrists think that they know the phenomenon of emotional lability which commonly accompanies brain damage. However, when we try to pin it down, it proves elusive. For example, one of emotionalism's most commonly quoted characteristics is its emotional inappropriateness, but this is denied by most patients (House *et al.* 1989). Perhaps its most striking and pathognomonic feature—the occasional coexistence of laughing and crying in the same patient—is also rare. Its only common and unequivocal feature is unheralded tearfulness which is socially embarrassing; a picture which is difficult to distinguish from the normal adjustment reaction in adults without brain damage. This creates a problem in recognizing pathological emotionalism in its milder forms, and knowing when poorly controlled emotional expression is plausibly attributable to the effects of brain damage rather than to the experience of ill health in general.

Apathy/emotional indifference

When Babinski (1914) described denial of hemiplegia after stroke and coined a term for it—anosognosia—he also described in the same patients a state of morbid emotional indifference to the illness from which they suffered. Although the term he proposed for that state—anosodiaphoria—was not adopted, the observation has stood the test of time. Unfortunately the two categories—verbal denial of illness and emotional indifference to illness—have been expanded to include a large number of disparate clinical phenomena, such as apathy, refusal to acknowledge a poor prognosis, and carelessness in modifying behaviour to take account of impairments (see e.g., Heilman *et al.* 1985).

Verbal denial and emotional indifference may accompany perceptual neglect or inattention, forming part of a cluster of phenomena discussed around the idea of lack of awareness of deficit. From the point of view of defining a syndrome, however, the problem is that each of these phenomena may occur independently of the others (Bisiach *et al.* 1986).

Apart from the problem of delineating the disorder, there is again a difficulty about ascribing it to brain pathology. If somebody says that they are aware of their condition and its prognosis, but they seem insufficiently upset by that knowledge, is that a manifestation of denial or morbid emotional indifference, or is it a manifestation of desirable coping responses such as stoicism?

'Orthodox' psychiatric syndromes

It is not unusual to encounter attempts to constrain the phenomena of brain damage within an orthodox psychiatric classification (for traumatic brain damage see, e.g., Lishman 1973; Grant and Alves 1987; for stroke see, e.g.,

Starkstein and Robinson 1989). This approach may be useful in some cases, but there is a danger that the standard categories may act as strait-jackets into which abnormal mood states in the brain-damaged are forced, simply because there is nowhere else to place them. For example, brain-damaged patients often have characteristics which could be construed as hypomanic —restlessness, over-talkativeness, unrealistic optimism, sexual disinhibition, insomnia. Is it justified to use the term hypomania to describe them? In the absence of more precise validation of our diagnostic criteria in brain damage, the answer must be—only with great caution and only speculatively. For that reason I think the concept of secondary or symptomatic mania (e.g., Krauthammer and Klerman 1978) needs to be viewed warily when applied to brain-damaged subjects (Starkstein *et al.* 1988).

One indication of the inadequacy of existing classifications is the degree to which they rely on terms such as 'organic psycho-syndrome' or 'organic personality change' as catch-alls. This approach is unfortunate because not only does it lump together changes which might be better dealt with separately, but it tends to emphasize stereotypes which can easily be thought of as inherent, biologically based, and unmodifiable.

Orthodox psychiatric classifications, which describe syndromes characterized by collections of certain affective, cognitive, and behavioural changes, will meet with difficulties even if we expand the syndromes used.

Table 9 The main features of emotional disorder encountered in brain-damaged adults

Affective	Behavioural	Cognitive	Somatic
Depression transient persistent	Tearfulness	Hopelessness Suicidal ideas	Anorexia and weight loss
Anxiety pervasive phobic	Reassurance-seeking	Worry Hypochondriasis	Insomnia Pain
Irritability	Non-compliance Aggression		Tension
Apathy	Inertia	Helplessness	Lethargy
Anhedonia	Social withdrawal		
Indifference	Carelessness	Down-playing	
Euphoria	Accident-proneness	or denial of disability/handicap	

The reason for this is that one of the consequences of brain damage is that it seems to lead to dissociation of functions which are normally integrated, so that partial or atypical syndromes abound. My own proposal is that we need (for now) to retreat from clinical typologies and start with careful descriptions of the basic symptoms and syndromes which are encountered in brain damage. The main ones are listed in Table 9. Interventions can then be targeted at each one individually without any prejudicial assumptions about aetiology or analogy with mood disorder in other populations.

Improving our measures of severity

Usually in psychiatry we define severity in terms of symptom levels, but which symptoms should we use? Many mood-rating scales score items which are not specific to brain damage. On the other hand, specific types of symptom (i.e., specific to brain damage) may reflect something that is qualitatively different about the person's mental state rather than quantitatively different. Is emotionalism associated with pathological laughter more severe or just different from emotionalism in which crying is the only phenomenon?

One illustration of the measurement problem is provided by a study using the Neurobehavioural Rating Scale, a 27-item scale which was derived from the Brief Psychiatric Rating Scale (BPRS) specifically for the assessment of brain-damaged adults (Levin *et al.* 1987). Emotional problems appeared unrelated to the severity of head injury as judged neurologically, but then so did a number of cognitive items like memory deficit. This raises the concern that the negative result could have been obtained because either the emotional items chosen were inappropriate, or the severity measure was too insensitive (see also Van Zomeren and Van den Berg 1985). There are numerous other rating scales which I will not review here, but to which the same concerns apply.

The alternative to measuring symptom levels is to measure disability or handicap. Undoubtedly, mood and handicap have an interrelationship, and we have all seen people whose major cause of social dysfunction is their emotional state. Unfortunately, handicap is over-determined and it is difficult to see how it could ever be used in a straightforward way as a measure of the severity of mood disturbance.

This problem of measurement is less of a disruption to clinical practice than some of the others I am discussing. My main purpose in raising it here is to act as a reminder that any intervention study must be assessed using a wide range of outcomes, even if the intervention is focused apparently on a specific phenomenon.

Improving our understanding of aetiology

We are likely to do better in treating symptoms in the brain-damaged if we know more about their causes.

As noted in the last section, it has proved consistently and frustratingly difficult to establish clear associations between the location of damage in the brain and the nature of emotional disorders which follow. It may be that the advent of more sophisticated techniques for brain imaging will improve the situation, but the search for brain behaviour correlates is always going to be difficult because the clinical causes of brain damage usually cross anatomical and functional boundaries. As a result, we remain uncertain about quite gross matters like whether the side of brain damage is important or whether there is a difference in the picture caused by predominantly subcortical as opposed to predominantly cortical damage.

Rather than review this literature, what I want to do is to mention an area of aetiology which is relatively neglected and yet which points to possibilities for treatment, and that is the role of interpersonal and social factors.

It is easy to neglect the dynamics of the situation in which brain-damaged people find themselves, because of the obvious structural damage which has been done to the brain. There are, of course, many psychological factors to explain why somebody with brain damage might develop a mood disorder. The trauma itself is a life event which results in many losses: of identity, of function (both physical and social) and of cherished plans. Major physical injury often leads to an enforced dependence on others so that the other social dimension which needs consideration is that of family dynamics. One illustration of the importance of social factors is that violence tends to be family-directed and may not remit over time (Brooks *et al.* 1986), unlike most physically induced pathologies. Expressed emotion is a concept which has proved useful in characterizing the impact of family relationships on other psychiatric disorders (Vaughn and Leff 1976). It is surprising that it has been neglected in this field.

The role of social and interpersonal context in explaining emotional problems could be explored in another direction. There may be social factors which put people at higher risk of acquiring brain damage when they are at the same time vulnerable to developing emotional problems. Vulnerability to traumatic brain damage is usually discussed in terms of individual characteristics (e.g., Sims 1985; Haas *et al.* 1987) but it may instead reflect the sensitivity of younger people to recent family changes. For example, there is evidence that accidents in children may be associated with antecedent stress for their mothers. By the same token, the sorts of life events which precede stroke are largely family-related experiences, such as bereavement or illness in relatives, which can in their own right lead to mood disorders (House *et al.* 1990).

Recent life events research suggests that it is circumstances where an event with particular significance is linked to a related difficulty which are especially likely to be associated with emotional problems (Brown and Harris 1989).

Improving our methods of clinical evaluation

There are three problems to be overcome in assessing patients with brain damage, two of which are not peculiar to the evaluation of mood. First, *distractibility* may lead to inconsistency in expressing and understanding emotion even when the latter is not the primary problem. Secondly, *socially unusual behaviour*, such as frequent swearing or touching, may make the verbal expression of other forms of emotion difficult to interpret because the way in which they are expressed is so unfamiliar to the interviewer.

The third difficulty in evaluation is posed by the fact that patients with brain damage often have *deficits in expressing and understanding emotion*. Emotion may be expressed by facial movement, by vocal intonation or by bodily movement and posture, and its expression in others may be read by attention to the same modalities. Brain damage has a number of non-specific motor effects which can impair expression of emotion. Such evidence as we have suggests that these effects lead patients to appear more depressed or apathetic than they feel, but that their perception of emotion in others is only as impaired as their general level of sensory deficit. However, it is a little-studied subject.

By contrast a lot of attention *has* been paid to disturbances of emotional expression and reception which arise from disruption of specific faculties. Disturbances of vocal expression and reception of emotion are now well recognized, and are usually known as aprosodias, that is, disturbances of speech prosody (Ross 1985). Disorders of facial expressiveness and of recognizing emotion in the faces of others are varieties of apraxia and agnosia, although they have not been given specific names (mercifully!).

In the face of these deficits, the use of observed behaviour and observed expression of emotion in the evaluation of mood will inevitably be more problematic than elsewhere in psychiatry. This is the most intractable of all the problems I am discussing; it has its major impact on the reliable implementation of any attempt to define or measure mood disorder and can only be minimized (not eliminated) by sensitive awareness of the presence of such deficits in the individual patient.

An outline of treatment approaches

Rather than review a series of non-definitive studies on various aspects of the treatment of emotional disorders in brain-damaged adults, I would like to outline the areas in which I see potential for intervention. The effectiveness of these interventions will only be demonstrated when we have resolved the problems of definition, measurement, and formulation outlined above, but in the meantime they should not be withheld simply for lack of rigorous experimental backing.

Counselling and related approaches

Because emotional disturbances are often thought of as meaningless in the context of brain damage, then quite simple techniques, such as counselling or grief work, are far too frequently neglected (Goodstein 1983). The teaching of problem-solving techniques is also neglected in my experience; perhaps because of an undue preoccupation with those patients who have no awareness that they have problems. A structured approach to such interventions can be made part of a physical rehabilitation programme (Oradei and Waite 1974; Goldberg *et al.* 1979).

Similarly, education is often neglected. Patients and their families may have gross misperceptions about the nature of cognitive and emotional problems following brain injury. In my experience, the two most common are: believing that peoples' personalities are permanently altered or 'lost' so that behaviour or emotional expression no longer means what it did; and being unaware of the distinction between focal and global intellectual deficits. A useful means of addressing these issues is during a clinical review of the results of neuropsychological testing or during a rehabilitation assessment, conducted with the patient or family.

Cognitive behaviour therapy

Most psychological therapy in the brain-damaged has been based on the idea that language-dependent therapies cannot be employed in this population. Particularly in the UK, the dominant model has been traditional behaviour therapy informed by standard reward–punishment paradigms and using, for example, the token economy (Power 1981; Wood 1987).

Certainly even severely damaged patients are sensitive to the non-verbal aspects of their milieu, as indicated by the context-dependence of much emotional disturbance. However, there are severe limitations to the behavioural approach, even if it could be shown that it was of any benefit for emotional problems. One obvious one is that it is rarely possible to control environmental contingencies in the sort of ward where most patients with brain damage are nursed.

In reality, there are very few patients who are so damaged that they cannot undertake some discussion of their situation, thus making them suitable candidates for a form of talking therapy. One example of the sort of treatment that might be useful is the cognitively oriented anger-control therapy which has been used in patients with impulse-control disorders in forensic psychiatry. Related ideas involve teaching patients to use distraction when they recognize themselves as entering situations which normally provoke emotional disturbance; for example, by turning to emotionally neutral topics when faced with the sort of stimulus which normally provokes emotionalism. I have certainly had some success with this technique. Other approaches which merit attention include exposure and desensitization, either directed at emotional expression itself or at other associated behaviour (Allman and House 1989).

Family-based interventions

Lezak (1978) has described some of the issues which arise during family interviews, and there are now several ways of standardizing both family assessment and intervention. The largely educational approach adopted in work on reducing expressed emotion in the families of schizophrenics would be one possible model (Leff *et al.* 1982). More formal family interventions, particularly based on the theories and practice of strategic family therapy, may well be appropriate in certain circumstances.

Much emotional disturbance arises as a response to circumstances rather than because of primarily psychological or interpersonal pathology. Families and patients need practical advice and support as well as therapy. Respite care provided by social services, family-placement schemes, or psychiatric admission, needs to be offered actively if it is to be taken up. Support groups like Headway can be helpful depending (as always) on local factors. With careful case management, resources can be 'networked' to produce a flexible support system for families (Rogers and Kreutzer 1984).

Physical methods of treatment

What about the physical methods of treatment which are used in other psychiatric disorders?

There is evidence that tricyclic antidepressants (Lipsey *et al.* 1984) and ECT (Murray *et al.* 1986) can be effective in treating depressive disorders in the brain-damaged, with the usual indications. Problems with side-effects may be considerable. For example, in one study using nortriptyline to treat depression in stroke patients approximately one-quarter had to be withdrawn because of adverse reactions (Lipsey *et al.* 1984). What is less clear is what other indications there may be for using psychotropic medication.

There are several reports which suggest that low dose antidepressants may be effective in treating emotionalism (Lawson and McLeod 1969; Schiffer *et al.* 1985). We are currently examining this possibility in a randomized prospective study which will look at the impact of mood disorder (defined in 'orthodox' psychiatric terms) and cognitive impairment on outcome of treatment aimed primarily at the emotionalism.

We need to know if there is a subgroup of patients with problems of indifference, apathy, withdrawal, or impaired arousal who might respond to more stimulating antidepressants or other psychotropics, such as protriptiline or tranylcypromine. If the newer generation of more specific serotonergic antidepressants have a role then a plausible hypothesis is that it may be in this group of conditions where arousal and attention deficits are major issues.

Benzodiazepines have all the drawbacks that characterize their general use. In addition, they may cause confusion and a sort of 'paradoxical' irritability and excitement. Anxiety or irritability may be better treated with low doses of major tranquillizers, especially those with fewer extrapyramidal side-effects, such as thioridazine and sulpiride. Sedating antidepressants, such as amitriptyline, can be useful in the management of anxiety and insomnia.

Other psychotropic drugs, such as lithium carbonate and carbamazepine, have had their exponents. Claims that anticonvulsants are especially useful in treating irritability or aggression are intriguing but unproven. As with the other treatments mentioned here, it is relatively easy to propose a feasible mechanism by which they might work and to make a case for evaluating their effectiveness in a formal treatment trial.

With any psychotropic medication, it is wise to start with low doses— which may be effective in themselves—and increase only slowly to the maximum tolerated.

Clinical guidelines

1. One must be prepared to accept as psychiatric problems states which are rare in other settings—such as emotionalism, marked emotional indifference or apathy. Even 'orthodox' psychiatric disorders, like agoraphobia, may look different in the context of brain damage. Clear delineation of the presenting problem is more help than attempting to use a standardized classification.

2. The diagnosis of personality change or organic personality disorder is counter-productive. It does not describe the problems, it distracts from making a wide-ranging aetiological formulation, and it suggests treatment will be ineffective.

3. Severity of emotional problems should be gauged by quantity of symptoms and their impact on function. Unusual symptoms like pathological laughter or complete denial of illness are not in their own right indications of severity.

4. An aetiological formulation should include consideration of the individual's social and family context: both before and since the injury.

The lessons learned from life events research and from characterizing family relationships can be imported with benefit. For that reason, family interviewing forms an important part of the assessment. Undertaken in the presence of the patient (however impaired), family interviewing can be a powerful means of helping all those present face and discuss the impact of a serious brain injury.

5. Basic psychological interventions are too frequently neglected. They should include education for patient and family, advice about rehabilitation and services, and counselling. Talking therapies may be effective, especially if problem-oriented. By comparison, the value of behaviour modification in routine clinical practice is often overestimated; in reality, it is difficult to deliver in the setting where brain-damaged patients are usually treated.

6. Involvement of the family should be a *sine qua non* of psychiatric intervention, as a minimum to educate, correct misapprehensions, and provide support.

7. The role of drugs is unclear. Where the usual indications for physical treatment exist—such as biological symptoms in depression—then they need not be withheld simply because of the different clinical context. Where the indications are uncertain, an individual therapeutic trial can be undertaken. Responses may be seen at lower dosage than in general psychiatric practice. In any case, dosage should start low and only build gradually because of the risk of adverse effects.

References

Allman, P. and House, A. (1989). Emotionalism: clinical features and management. *Geriatric Medicine*, **20**, 43–8.

Babinski, M. J. (1914). Contribution a l'etude des troubles mentaux dans l'hemiplegie organique cerebrale. *Revue Neurologique*, **27**, 845–8.

Bisiach, E., Vallar, G., Perani, D., Papagno, C., and Berti, A. (1986). Unawareness of disease following lesions of the right hemisphere: anosognosia for hemiplegia and anosognosia for hemianopia. *Neuropsychologia*, **24**, 471–82.

Brooks, N., Campsie, L., Symington, C., Beattie, A., and McKinlay, W. (1986). The 5 year outcome of severe blunt head injury: a relative's view. *Journal of Neurology, Neurosurgery and Psychiatry*, **49**, 765–70.

Brown, G. and Harris, T. (1989). *Life events and illness*. Guilford, New York.

Goldberg, R. L., Wise, T. N., and Le Buffe, F. P. (1979). The stroke unit: psychological aspects of recovery. *Psychosomatics*, **20**, 316–21.

Goodstein, R. K. (1983). Overview: cerebrovascular accident and the hospitalised elderly—a multidisciplinary approach. *American Journal of Psychiatry*, **140**, 141–7.

Grant, I. and Alves, W. (1987). Psychiatric and psychosocial disturbances in head injury. In *Neurobehavioural recovery from head injury* (ed. H. S. Levin, J. Grafman, and H. M. Eisenberg), pp. 232–61. Oxford University Press.

Haas, J., Cope, D., and Hall, K. (1987). Premorbid prevalence of poor academic performance in severe head injury. *Journal of Neurology, Neurosurgery and Psychiatry*, **50**, 52–6.

Heilman, K., Bowers, D., and Valenstein, E. (1985). Emotional disorder associated with neurological disease. In *Clinical neuropsychology* (2nd edn) (ed. K. M. Heilman and E. Valenstein), pp. 377–402. Oxford University Press.

House, A., Dennis, M., Warlow, C., and Hawton, K. (1989). Emotionalism after stroke. *British Medical Journal*, **298**, 991–4.

House, A., Dennis, M., Mogridge, L., Hawton, K., and Warlow, C. (1990). Life events and difficulties preceding stroke. *Journal of Neurology, Neurosurgery and Psychiatry*, **53**, 1024–8.

Krauthammer, C. and Klerman, G. (1978). Secondary mania: manic syndromes associated with antecedent physical illness or drugs. *Archives of General Psychiatry*, **35**, 1333–9.

Lawson, I. R. and McLeod, R. D. (1969). The use of imipramine and other psychotropic drugs in organic emotionalism. *British Journal of Psychiatry*, **115**, 281–5.

Leff, J., Kuipers, L., Berkowitz, R., Eberlien-Vries, R., and Sturgeon, D. (1982). A controlled trial of social intervention in the families of schizophrenic patients. *British Journal of Psychiatry*, **141**, 121–34.

Levin, H. S., *et al.* (1987). The neurobehavioural rating scale: assessment of the behavioural sequelae of head injury by the clinician. *Journal of Neurology, Neurosurgery and Psychiatry*, **50**, 183–93.

Lezak, M. (1978). Living with the characterologically altered brain injured patient. *Journal of Clinical Psychiatry*, **39**, 592–8.

Lipsey, J., Robinson, R., Rao, K., and Price, T. (1984). Nortriptyline treatment of poststroke depression. *Lancet*, **i**, 297–300.

Lishman, W. A. (1973). The psychiatric sequelae of head injury: a review. *Psychological Medicine*, **3**, 304–18.

Murray, G., Shea, V., and Conn, D., (1986). Electroconvulsive therapy for poststroke depression. *Journal of Clinical Psychiatry*, **47**, 258–60.

Oradei, D. and Waite, N. (1974). Group psychotherapy with stroke patients during the immediate recovery period. *American Journal of Orthopsychiatry*, **44**, 386–95.

Power, G. E. (1981). *Brain function therapy*. Gower, London.

Rogers, P. and Kreutzer, J. (1984). Family crises following head injury: a network intervention strategy. *Journal of Neurosurgical Nursing*, **16**, 343–6.

Ross, E. D. (1985). Modulation of affect and non-verbal communication by the right hemisphere. In *Principles of behavioural neurology* (ed. M-M. Mesulam), pp. 239–57. Davis, Philadelphia.

Schiffer, R., Herndon, R., and Rudick, R. (1985). Treatment of pathologic laughing and weeping with amitriptyline. *New England Journal of Medicine*, **312**, 1480–2.

Sims, A. C. (1985). Head injury, neurosis and accident proneness. *Advances in Psychosomatic Medicine*, **13**, 49–70.

Starkstein, S. and Robinson, R. (1989). Affective disorders and cerebral vascular disease. *British Journal of Psychiatry*, **154**, 170–82.

Starkstein, S., Boston, J., and Robinson, R. (1988). Mechanisms of mania after brain injury. *Journal of Nervous and Mental Diseases*, **176**, 87–100.

Van Zomeren, A. H. and Van den Berg, W. (1985). Residual complaints of patients 2 years after severe head injury. *Journal of Neurology, Neurosurgery and Psychiatry*, **48**, 21–8.

Vaughn, C. and Leff, J. (1976). The influence of family and social factors on the course of psychiatric illness. A comparison of schizophrenic and depressed neurotic patients. *British Journal of Psychiatry*, **129**, 125–37.

Wood, R. L. (1987). *Brain injury rehabilitation: a neurobehavioural approach*. Croom Helm, London.

6

Serotonin re-uptake inhibitors in obsessive–compulsive disorder: what is their therapeutic role?

JOHN COBB

Introduction of clomipramine

Iridescent green liquid passing slowly through a drip is an enduring memory of the first patient I encountered undergoing treatment for obsessive–compulsive disorder (OCD). The danger of producing haemoglobinuria, cardiac arrhythmias, and epileptic seizures, together with the inevitable venous inflammation, heightened the drama of an occasion which appeared to an inexperienced student to be more Gothic than scientific. The liquid contained clomipramine, a substance which after the first claim of its effectiveness in OCD (Fernandez and Lopez-Ibor 1969), rapidly gained the reputation of having specific 'anti-obsessional' properties (Capstick 1975). In retrospect it is salutary that therapeutic enthusiasm outstripped scientific caution in a rush to treat a previously disabling but intractable condition. Thus, intravenous use was advocated despite good evidence that clomipramine was perfectly well absorbed after oral administration (Beaumont 1973). The reputation clomipramine acquired as an intravenous treatment carried over to its use in capsule form, and it continued to be seen by many clinicians as having unique properties (Cobb 1977).

Pharmacologically, clomipramine is distinct from other tricyclics in its effect on the re-uptake of transmitters at pre-synaptic terminals. Ranged on a spectrum, clomipramine has a powerful inhibitory effect on serotonin (5-HT) re-uptake and a weak effect on noradrenaline. In contrast, amitriptyline has an equal effect on both, whereas desipramine largely inhibits noradrenaline and has little effect on serotonin (Horn 1976). Recognition of special neuropharmacological properties, together with considerable evidence of its efficacy from clinical reports and open studies, set the scene for controlled clinical trials of clomipramine and, subsequently, more specific newer serotonergic drugs in OCD. Before these are reviewed it is necessary to look at some of the particular pitfalls encountered in research in this area. Ignorance of these leads to bias or confusion in the interpretation of results.

Controversial aspects of obsessive–compulsive disorder (OCD)

Prevalence and incidence

Until recently, OCD was believed to be relatively uncommon. Among psychiatric populations incidence ranged from 0.1 to 4.6 per cent (Black 1974). As a rule of thumb, most practising psychiatrists in the UK reckon on seeing one new case with this disorder each year. Population studies estimated prevalence rates at around 0.05 per cent (Black 1974). In practice this meant that even in centres specializing in the disorder, recruitment of suitable patients for a controlled study took a long time and even then the total number tended to be low. Above 10 patients in each treatment cell represented a considerable achievement (Marks *et al.* 1980).

Recent general population surveys, however, have reported staggeringly higher prevalence rates, ranging from 2 to 3 per cent (Robins *et al.* 1984; Karno *et al.* 1987). Intriguingly, patients seem to have been easier to find for controlled studies, because no fewer than 15 were reported between 1985 and 1990, compared to five between 1980 and 1985. On the one hand, OCD patients are well known to be secretive concerning their symptoms. A medical practitioner whose wife had for some years spent two hours each morning and evening in the bathroom was unaware that her symptoms were caused by obsessive neurosis. Previous community surveys may well have grossly underestimated the prevalence. On the other hand, obsessional symptoms are common features of a primary depressive illness. Recent figures are based on the Diagnostic Interview Schedule given by non-medical personnel. This does not allow for evaluation of the key issue of when obsessional symptoms are adequate to substantiate a diagnosis of OCD rather than forming part of some other disorder (Bebbington 1990). By contrast, the Present State Examination, which allows the interviewer considerable latitude in not diagnosing OCD when symptoms occur in relative isolation, reveals much lower prevalence rates of around 0.2 per cent (Bebbington 1990).

Association of OCD and anxiety

Both the DSM-III-R and ICD diagnostic schemes classify OCD under anxiety disorders, implying similarities in psychopathology. An increasing body of evidence argues that this is inappropriate (Montgomery 1990):

1. symptom complexes in OCD are distinctive;
2. OCD sufferers have a long, chronic, fluctuating course;
3. the diagnosis of OCD has a high stability over time, unlike anxiety states which often develop into depression;
4. OCD has an equal sex distribution, whereas in anxiety disorders females outnumber males by 2:1;
5. anxiety states have an earlier age of onset than OCD;

6. OCD does not respond to anxiolytics, such as benzodiazepines, or to psychological methods of anxiety treatment, such as relaxation; and
7. in contrast to sufferers from anxiety, patients with OCD respond poorly to placebo.

Most studies report a placebo response of under 20 per cent. Thus, although patients with OCD often show high levels of anxiety it is unjustifiable and misleading to view this anxiety as a primary feature of the psychopathology.

Association of OCD and depression

Patients with OCD often suffer from significant depressive symptomatology. The depressed group can be further divided into those who are depressed as a result of their obsessional symptoms, and those who have been misdiagnosed and are really suffering from a primary depression. Tormented by disturbing and incessant thoughts which force the patient to carry out absurd, time-consuming rituals, it is hardly surprising that a secondary depression develops. Normal life, although longed for, seems to be hopelessly out of reach. Differentiation of such a state from a primary depression which includes obsessional features among its symptomatology is mainly based on the clinical history, and is easier to achieve in theory than in practice. In the large epidemiological catchment studies (Karno *et al.* 1987) one-third of patients with OCD also fulfilled DSM-III-R criteria for major depression. Failure in achieving a distinction underlies the controversy concerning whether clomipramine's apparent success in OCD is due to its effectiveness as an antidepressant or reflects a separate 'anti-obsessional' activity. Inclusion of patients with primary depression, misdiagnosed as OCD, will bias the outcome of a controlled treatment study, particularly if the overall numbers are low.

Nature of the population studied

Patients with OCD tend to be treated in specialist centres and these centres in their turn tend to attract referral of the type of patient likely to respond to the methods for which a centre is famous. Thus, when a centre with an international reputation for the quality of its behavioural therapy (Marks *et al.* 1980, 1988) obtains less impressive results with clomipramine than several other centres with a pharmacological bias, one is not surprised. The difference may be more with the type of patient treated than with the treatment method employed.

Studies with clomipramine

Placebo-controlled studies

A total of 11 double-blind studies comparing clomipramine with placebo have been reported (Thoren *et al.* 1980; Marks *et al.* 1980; Montgomery

1980; Insel *et al*. 1983; Mavisskalian *et al*. 1985; Flament *et al*. 1985; Marks *et al*. 1988; Benkelfat *et al*. 1989; Jenike *et al*. 1989; De Veaugh-Geiss *et al*. 1989; Greist 1989). One of these was a large multi-centre investigation involving 384 patients (De Veaugh-Geiss *et al*. 1989). The others were all single-centre projects involving small cohorts of patients ranging from 12 to 49 in number. All the studies attempted to exclude patients with a primary diagnosis of depression, and five distinguished between OCD patients with and without secondary depression. Four of the earlier studies ran treatment for less than six weeks. Unlike the treatment of affective disorders, full response of OCD to medication has been shown in a number of studies to take up to eight weeks, so that these studies may not have allowed sufficient time for the treatment results to become apparent. Despite this and the small numbers involved, all the studies except two (Marks *et al*. 1980, 1988) showed an impressive statistically significant advantage of clomipramine over placebo. Five out of six studies showed that the significant superiority of clomipramine extended to non-depressed patients.

The large multi-centre study (De Veaugh-Geiss *et al*. 1989) judged that just over 50 per cent of patients treated could be classified as 'responders'. Patients who responded were either free of symptoms (a relatively small group), or else were so much improved that residual symptoms did not interfere significantly in their lives. Results of this magnitude were typical of the studies as a whole. Placebo responses were uncommon, being less than 20 per cent overall.

In double-blind placebo controlled studies, 9 out of 11 positive outcomes should be enough to satisfy most clinicians of the anti-obsessional properties of clomipramine. However, the two exceptions (Marks *et al*. 1980, 1988) cannot be dismissed lightly. With the exception of the multi-centre study, these are by far the largest studies, including 37 and 49 patients, respectively. They were carried out at the Maudsley Hospital, a centre renowned for its diagnostic rigour. Apart from one other small ($n = 14$) study (Montgomery 1980), all the remaining studies were carried out in the US. Under the leadership of Professor Marks, the Maudsley Group has played a leading role in the development of behavioural therapy treatment for a range of disorders, including OCD. Both studies there were based on a 2×2 design to enable the evaluation of both behavioural treatment and clomipramine with placebo. In the first study, exposure and response prevention were compared to a placebo treatment. In the second, exposure was compared to 'anti-exposure'. Both studies compared clomipramine with placebo. Results from the first study showed a clomipramine effect, but this was judged to be only on the 'depressive' symptoms and not on the 'obsessional' symptoms. Interpretation of the results led to difficulties and one of the senior collaborators in the study published results elsewhere (Rachman *et al*. 1979) pointing out that as the major measures were carried out 7 weeks after the start of treatment, insufficient time may have elapsed to allow a 'clomipramine effect' to be apparent.

The subsequent project (Marks *et al.* 1988) studied non-depressed patients. Clomipramine was found to be superior at 8 weeks, but this superiority did not persist to the end of treatment at 27 weeks. Exposure *in vivo* and response prevention were shown in both studies to have a powerful therapeutic effect. However, the use of 'anti-exposure' as a controlled procedure must be queried. In another study (Cottraux *et al.* 1990), patients complied very poorly with 'anti-exposure', presumably because for them it lacked face validity. The general public are increasingly aware of the value of behavioural techniques. There is widespread coverage in newspapers and magazines; approximately five million people in the UK recently watched a 45 minute BBC programme on the successful behavioural treatment of a patient with OCD. Both in the US and the UK *The boy who couldn't stop washing* (Rapoport 1990) has become a bestseller. With this information, whether asked to or not, patients are likely to practise behaviourally based strategies, particularly if they are already starting to feel that both their drive to carry out rituals and the strength of their compulsions are starting to weaken. The Maudsley results can thus be interpreted as validating behavioural therapy as an alternative to clomipramine or supporting the combination of clomipramine and behavioural therapy.

Clomipramine and other tricyclics

If it is established that clomipramine has an effect on OCD, the next question concerns the mechanism. Though itself a powerful inhibitor of pre-synaptic 5-HT uptake, clomipramine's major metabolite, desmethyl-clomipramine (DMCP), is a potent noradrenaline re-uptake blocker. Some studies (e.g., Mavissakalian 1990; Stern *et al.* 1980) have shown a correlation between plasma clomipramine levels and response of obsessional symptoms, but no correlation with plasma DMCP levels. Other studies have failed to replicate this (e.g., Kasvikis and Marks 1988).

A number of studies have compared clomipramine with other tricyclics. Tricyclics vary according to the ratio of effects on noradrenaline compared to effects on 5-HT. Two double-blind studies, one in children (Leonard *et al.* 1988) and one in adults (Zohar *et al.* 1988), have shown clomipramine to be significantly superior to other tricyclics, with desipramine, a selective noradrenaline uptake inhibitor, having little or no effect on obsessive symptoms. Three double-blind studies (Ananth *et al.* 1981; Volavka *et al.* 1985; Foa *et al.* 1987) comparing clomipramine with tricyclics with mixed and fairly similar effects on noradrenaline and 5-HT (amitriptyline and imipramine) produced similar results, whereas one study involving imipramine (Thoren *et al.* 1980) showed only a weak superiority for clomipramine. These studies further strengthen the hypothesis that the anti-obsessional efficacy of clomipramine is related specifically to its serotonin re-uptake blocking properties.

New 5-HT re-uptake blockers

Efficacy

Patients with OCD are, either by nature or as a consequence of their illness, meticulous when it comes to tablet-taking. Thus, most studies show little difference in drop-out rates between those on clomipramine and those on placebo. Clomipramine produces more severe side-effects than other tricyclics (see Table 3, p. 23), and overdose with clomipramine can be fatal. Development of a range of non-tricyclic serotonin re-uptake blockers with antidepressant effect equivalent to that produced by standard tricyclics (see Chapter 3) provided the opportunity for testing the hypothesis that selective 5-HT re-uptake blockers may have anti-obsessional actions.

So far, four such drugs have been released in the UK: fluvoxamine, fluoxetine, sertraline, and paroxetine. Although differing markedly in chemical structure and in some significant pharmacological properties they resemble one another in having little or no affinity for serotonin, dopamine, adrenergic, histaminic, or muscarinic receptors, their effect *in vivo* being largely confined to the pre-synaptic blockade of serotonin re-uptake, with only minimal effect on the re-uptake of dopamine and noradrenaline. Clinically, they are largely devoid of the anti-cholinergic side-effects of the tricyclics (see Chapter 3).

Several open studies have indicated that fluvoxamine (Price *et al.* 1987) and fluoxetine (Turner *et al.* 1985; Fontaine and Chournard 1985) are effective in the treatment of OCD. Four double-blind placebo controlled studies have shown fluvoxamine to be significantly superior to placebo (Perse *et al.* 1987; Goodman *et al.* 1989; Jenike *et al.* 1990*a*; Cottraux *et al.* 1990). One of these studies (Perse *et al.* 1987) used a cross-over design on a small group of patients ($n = 16$). The others, two carried out in different centres in the UK and US, and one in France, used parallel group design with a population of around 40 in each study. All studies found fluvoxamine to be effective, irrespective of whether or not the obsessional symptoms were accompanied by depression.

One study (Cottraux *et al.* 1990) combined drug and behavioural therapy in a design similar to the Marks *et al.* (1980, 1988) studies described earlier. However, this study compared fluvoxamine with fluvoxamine plus anti-exposure and exposure plus placebo. Thus there was no true placebo group. Cottraux *et al.* (1990) found fluvoxamine to be superior to exposure plus placebo at eight weeks although this superiority was subsequently lost.

Goodman *et al.* (1989) measured response on a weekly basis. Improvement began at the end of the first week, but did not become significant until treatment had been continued for at least six weeks, patients continuing to improve for at least two weeks after that. Not only is this pattern of improvement somewhat different from that shown in depressed patients, but it also emphasizes the importance of continuing treatment for a full eight weeks before discarding it as ineffective. Nine out of 21 patients on fluvoxamine were 'much improved', compared to none of the 21 on placebo. One-half of the patients in each group were depressed and it was shown that prognosis with treatment was independent of initial levels of depression and of severity of obsessional symptoms. It was noted that in some patients depression improved but not obsessional symptoms. Treatment was well tolerated and few patients were troubled by side-effects. Improvement in obsessions preceded that in compulsions, although by the end of the study both showed similar gains. This contrasts with the results from behavioural therapy which is known to be much more effective for rituals than ruminations (Foa *et al.* 1985).

Side-effects and toxicity

Side effects and possible toxicity are a major concern to clinicians when a new group of potentially useful drugs is introduced. Zimelidine, a bicyclic compound, was a 5-HT uptake blocker which was shown to be an effective antidepressant. Three out of four studies showed that it had a significant effect in OCD (Fontaine *et al.* 1985; Kahn *et al.* 1984; Insel *et al.* 1985; Prasad 1984). Unfortunately, just at the time that it was gaining a sound clinical reputation, it had to be withdrawn because of an infrequent but serious toxic reaction. Both fluvoxamine and fluoxetine have now been widely used for over five years. Preliminary reports have warned that fluoxetine may have an activating effect during the first two weeks of treatment, which in depressed patients may be associated with an increase in suicidal thoughts (p. 27). There have also been reports of a rare but fatal systemic vasculitis associated with fluoxetine which led to the recommendation that fluoxetine be withdrawn if a rash develops.

Preliminary reports of controlled studies involving sertraline (Bick and Gaffney in press; Chouinard *et al.* 1990) indicate that this is also significantly more effective than placebo in OCD, although there are early findings which suggest that the therapeutic benefit is not as great as that produced by fluvoxamine (Jenike *et al.* 1990*b*), possibly indicating that very high 5-HT uptake selectivity weakens a drug's effectiveness in OCD. OCD can be divided into subtypes on clinical grounds. However, it has not been possible to identify biological or clinical factors which will predict response to medication (Goodman *et al.* 1989), although patients with schizotypal personality may have a poor outcome (Jenike *et al.* 1986).

Drug combinations in OCD

About a half of patients 'much improved' with fluvoxamine (Goodman *et al.* 1989) or clomipramine (Greist 1989) leaves half either 'improved' (still with significant symptoms) or unchanged. An uncontrolled study suggests that such resistant patients may be helped by augmenting clomipramine treatment with lithium (Rasmussen 1984) or L-tryptophan (Rasmussen 1984; Yaryura-Tobias and Bhagavan 1977). Recently, however, 17 patients who had failed to respond to at least eight weeks of treatment with fluvoxamine either with or without lithium were reported in an open case series (McDougle *et al.* 1990). Of these patients, 50 per cent showed worthwhile improvement when a neuroleptic (either thioridazine or pimozide) was added. Patients with either schizotypal personalities or associated Tourette tic disorder were particularly likely to respond. Markovitz *et al.* (1990) found a similar improvement when patients who had failed to respond to fluoxetine had their treatment augmented with buspirone.

Role of behavioural therapy

Turning to the question of the effectiveness of behavioural therapy compared to medication, a comprehensive review of results of studies of behavioural therapy carried out in various countries involving over 200 patients (Foa *et al.* 1985) produced results which were remarkably similar to those from drug studies. That is, 51 per cent with reduction in symptoms of at least 70 per cent (much improved or cured), 39 per cent with reductions ranging from 31 to 69 per cent (improved), and the rest unchanged. Although impressive, it is important to remember that with either drugs or behavioural therapy a substantial minority are left with significant difficulties. Whether the patients who respond to drugs are also those who would respond to behavioural therapy is a question which remains unanswered. Certainly, neither of the studies which combined behavioural therapy with clomipramine (Marks *et al.* 1988) or fluvoxamine (Cottraux *et al.* 1990) showed any advantage of combined treatment over behavioural therapy alone, but then in both studies behavioural therapy had a powerful effect and, as already noted, the 'placebo' behavioural therapy treatments can be questioned. Patients are likely to practise 'behavioural therapy', whether instructed by therapists or not. The importance of this is heightened when considering the problem of relapse. Early studies involving medication alone and studies in which behavioural therapy has been specifically excluded showed very high rates of relapse when the drugs were discontinued (Griest 1990; Pato *et al.* 1984). In contrast, follow-up studies of up to five years following successful behavioural treatment show that improvement holds up extremely well, with 76 per cent of patients retaining improvement after a mean of one year follow-up (Foa *et al.* 1985). More recent studies of 'drug alone' and studies when drugs have been combined with behavioural therapy yield much lower relapse rates after medication is discontinued (Fontaine and Chouinard 1985; Marks *et al.* 1988). As behavioural methods are more widely known, it is likely that patients practise exposure and response prevention by themselves, once improved by the medication. This tendency should probably be encouraged by taking a combined pharmacological and psychological approach to treatment (Greist 1989).

Mechanism of drug effects

A guiding concept in the search for new psychotherapeutic agents has been the increasing selectivity of drugs for their actions on single neurotransmitter systems. Although this has led to definite progress in the understanding and treatment of OCD, it is most unlikely that the disorder will ever be explained on the basis of an abnormality of a single neurotransmitter system.

Clinical experience and the fact that treatments which are highly effective in certain patients are ineffective in others, makes one wary of the idea that any single hypothesis could ever account for the diverse, complex symptomatology of OCD. For example, although controlled studies with monoamine oxidase inhibitors (MAOIs) (Insel *et al.* 1983) have produced conflicting results, I have seen convincing clinical improvement with both phenelzine and tranylcypromine in two patients who have failed to respond to either clomipramine or behavioural therapy.

An unanswered question concerns the mechanism of action of drug treatment. Inhibition of 5-HT re-uptake occurs immediately, although clinical improvement is not apparent for several weeks. Obviously, early serotonergic effects are only the beginning of a complex chain of neurophysiological and psychological changes leading to clinical change. Even though neuropsychological abnormalities clearly play an important role in OCD, day-to-day management of patients reminds one of the importance of cognitive (Salkovskis and Kirk 1989), interpersonal, and psychosocial factors.

Clinical guidelines

Bearing in mind that one must be cautious in inferring pathogenesis from pharmaceutical responsiveness, there is little doubt that work involving serotonin re-uptake blocking drugs has advanced our understanding of OCD and our ability to treat what was traditionally known as a disabling and intractable disorder.

Having been involved in the assessment and management of over 250 patients with OCD during the past 17 years, I would recommend the following specific management plan in addition to a general psychiatric approach to personal, family and social problems.

1. Treatment should begin with an initial course of either fluvoxamine in doses increasing up to 300 mg daily or of clomipramine, again increasing to 300 mg if this can be tolerated. Clinicians used to treating depression need to be reminded that response takes at least 8 weeks. Although both drugs are equally effective the lower incidence of side-effects has recently led me to change to fluvoxamine as the drug of choice. Along with this, patients should be encouraged to adopt sound behavioural strategies when facing their problems and in severe cases provided with a self-help manual (Marks 1981; Greist 1990).

2. Patients often come for treatment with dogmatic attitudes concerning what treatments are acceptable. For a variety of reasons, about 25 per cent decline behavioural treatment (Foa *et al.* 1985; McDonald *et al.* 1988) and probably a similar proportion are biased against drugs. Although one can use discussion, sharpened by cognitive techniques, to challenge irrational beliefs concerning treatment, it makes sense when we have the option of two proven effective treatments to give patients what they want. Thus the sizeable group of patients who are opposed to the idea of medication should be offered behavioural therapy immediately after assessment. Even though specialist skills are required, the time involved may be much less than formerly believed. Marks *et al.* (1988) produced very good results with only three hours of skilled therapist time.

3. For patients who do not improve with 8 weeks of medication alone, behavioural therapy should be added. Up to 20 hours of therapist accompanied exposure and response prevention may be required.

4. Failure to respond may also be managed by changing from one drug to another (e.g., fluvoxamine to clomipramine) or, after a suitable wash-out period, by trying an MAOI (tranylcypromine or phenelzine). Combinations with lithium and neuroleptics should also be tried.

5. In the very small group of chronic, severely disabled patients who fail to respond to the above measures, stereotactic limbic leucotomy should be considered (Cobb and Kelly 1990).

References

Ananth, J., Pecknold, J. C., Van Den Steen, N., and Engelsmann, F. (1981). Double-blind comparative study of clomipramine and amitriptyline in obsessive neurosis. *Progress in Neuropsychopharmacology, Biology and Psychiatry*, **5**, 257–62.

Beaumont, G. (1973). Clomipramine in the treatment of obsessive-compulsive disorder. *Journal of International Medical Research*, **1**, 423–4.

Bebbington, P. (1990). The prevalence of OCD in the community. In *Obsessive compulsive disorder* (ed. S. A. Montgomery, W. K. Goodman, and N. Goeting), pp. 7–18. Duphar, Southampton.

Benkelfat, C., Murphy, D. L., Zohar, J., Hill, J. L., Grover, G., and Insel, T. R. (1989). Clomipramine in obsessive-compulsive disorder. Further evidence for a serotonergic mechanism of action. *Archives of General Psychiatry*, **46**, 23–8.

Bick, D. A. and Gaffney, M. (in press). Results of a double-blind placebo controlled trial using a new serotonin reuptake inhibitor, sertraline, in obsessive-compulsive disorder. *Psychopharmacology Bulletin*.

Black, A. (1974). The natural history of obsessional neurosis. In *Obsessional states* (ed. H. R. Beech), pp. 19–54. Methuen, London.

Capstick, N. (1975). Clomipramine in the treatment of the true obsessional state—a report on four patients. *Psychosomatics*, **16**, 21–5.

Chouinard, G., *et al.* (1990). Results of a double-blind placebo controlled trial of a new serotonin uptake inhibitor, sertraline, in the treatment of obsessive-compulsive disorder. *Psychopharmacology Bulletin*, **26**, 279–84.

Cobb, J. (1977). Drugs in treatment of obsessional and phobic disorders with behavioural therapy—possible synergism. In *The treatment of phobic and obsessive compulsive disorders* (ed. Boulougouris, J. C. and Rabivilas, A.), pp. 127–38. Pergamon, Oxford.

Cobb, J. and Kelly, D. (1990). Psychosurgery: is it ever justified? In *Dilemmas and difficulties in the management of psychiatric patients* (ed. K. Hawton and P. Cowen), pp. 219–30. Oxford University Press.

Cottraux, J., *et al.* (1990). A controlled study of fluvoxamine and exposure in obsessive-compulsive disorder. *International Clinical Psychopharmacology*, **5**, 17–30.

De Veaugh-Geiss, J., Landau, P., and Katz, R. (1989). Treatment of obsessive compulsive disorder with clomipramine. *Psychiatric Annals*, **19**, 97–101.

Fernandez, R. and Lopez-Ibor, R. (1969). Mono-chlorimipramine in the treatment of psychiatric patients resistant to other therapies. *Actas Luso Españolas de Neurologias y Psiquiatria*, **26**, 119–47.

Flament, M. F., *et al.* (1985). Clomipramine treatment of childhood obsessive compulsive disorder. A double-blind controlled study. *Archives of General Psychiatry*, **42**, 977–83.

Foa, E. B., Steketee, G. S., and Ozarow, B. J. (1985). Behavior therapy with obsessive-compulsives from theory to treatment. In *Obsessive-compulsive disorder: psychological and pharmacological treatment* (ed. M. Mavissakalian, S. M. Turner, and L. Michelson), pp. 49–129. Plenum, New York.

Foa, E. B., Steketee, G., Kozak, M. J., and Dugger, D. (1987). Effects of imipramine on depression and obsessive-compulsive symptoms. *Psychiatry Research*, **21**, 123–36.

Fontaine, R. and Chouinard, G. (1985). Fluoxetine in the treatment of obsessive compulsive disorder. *Progress in Neuropsychopharmacology and Biological Psychiatry*, **9**, 605–8.

Fontaine, R., Chouinard, G., and Iny, L. (1985). An open clinical trial of zimeldine in the treatment of obsessive compulsive disorder. *Current Therapeutic Research*, **37**, 326–32.

Goodman, W. K., Price, L. H., Rasmussen, S. A., Delgado, P. L., Heininger, G. R., and Charney, D. S. (1989). Efficacy of fluvoxamine in obsessive-compulsive disorder. *Archives of General Psychiatry*, **46**, 36–44.

Goodman, W. K., et al. (1990). Specificity of serotonin reuptake inhibitors in the treatment of obsessive-compulsive disorder. *Archives of General Psychiatry*, **47**, 577–85.

Greist, J. H. (1989). *Obsessive-compulsive disorder: a guide*. Anxiety Disorders and Lithium Information Centers, Madison, Wis.

Greist, J. H., Jefferson, J. W., Rosenfeld, R., Gutzmann, L. D., March, J. S., and Barklage, N. (1990). Clomipramine and obsessive compulsive disorder. A placebo controlled double-blind study of 32 patients. *Journal of Clinical Psychiatry*, **51**, 292–7.

Horn, A. S. (1976). The interaction of tricyclic antidepressants with the biogenic amine uptake systems in the central nervous system. *Postgraduate Medical Journal*, **52**(Suppl. 3), 25–31.

Insel, T. R., Murphy, D. L., Cohen, R. M., Alterman, I., Kitts, C., and Linnoila, M. (1983). Obsessive-compulsive disorder: a double-blind trial of clomipramine and clorgyline. *Archives of General Psychiatry*, **40**, 605–12.

Insel, T. R., Mueller, E. A., Alterman, I., Linnoila, M., and Murphy, D. L. (1985). Obsessive-compulsive disorder and serotonin: is there a connection? *Biology of Psychiatry*, **20**, 1174–88.

Jenike, M. A., Baer, L., Minichiello, W. E., Schwartz, C. E., and Carey, R. J. (1986). Concomitant obsessive-compulsive disorder and schizotypal personality disorder. *American Journal of Psychiatry*, **143**, 530–2.

Jenike, M. A., Baer, L., Summergrad, P., Weilburg, J. B., Holland, A., and Seymour, R. (1989). Obsessive-compulsive disorder: a double-blind, placebo-controlled trial of clomipramine in 27 patients. *American Journal of Psychiatry*, **146**, 1328–30.

Jenike, M. A., et al. (1990a). A controlled trial of fluvoxamine in obsessive-compulsive disorder: implications for a serotonergic theory. *American Journal of Psychiatry*, **147**, 1209–15.

Jenike, M. A., Baer, L., Summergrad, P., Minichiello, W. E., Holland, A., and Seymour, R. (1990b). Sertraline in obsessive-compulsive disorder: a double-blind comparison with placebo. *American Journal of Psychiatry*, **147**, 923–8.

Kahn, R. S., Westenberg, H. G. M., and Jolies, J. (1984). Zimelidine treatment of obsessive-compulsive diosrder. *Acta Psychiatrica Scandinavica*, **69**, 259–61.

Karno, M. et al. (1987). Lifetime prevalence of specific psychiatric disorders among Mexican Americans and non-Hispanic whites in Los Angeles. *Archives of General Psychiatry*, **44**, 695–701.

Kasvikis, Y. and Marks, I. M. (1988). Clomipramine in obsessive-compulsive ritualisers treated with exposure therapy: relations between dose, plasma levels, outcome and side-effects. *Psychopharmacology*, **95**, 113–18.

Leonard, H., Swedo, S., Rapoport, J. L., Coffey, M., and Cheslow, D. (1988). Treatment of childhood obsessive-compulsive disorder with clomipramine and desmethyl-imipramine: a double-blind crossover comparison. *Psychopharmacology Bulletin*, **24**, 93–5.

Markovitz, P. J., Stagno, S. J., and Calabrese, J. R. (1990). Buspirone augmentation of fluoxetine in obsessive-compulsive disorder. *American Journal of Psychiatry*, **147**, 798–800.

Marks, I. M. (1981). *Cure and care of neurosis*. John Wiley, New York.

Marks, I. M., Stern, R. S., Mawson, C., Cobb, J., and McDonald, R. (1980). Clomipramine and exposure for obsessive compulsive rituals. *British Journal of Psychiatry*, **136**, 1–25.

Marks, I., *et al.* (1988). Clomipramine, self-exposure and therapist aided exposure in obsessive-compulsive ritualisers. *British Journal of Psychiatry*, **152**, 522–34.

Mavissakalian, M., Turner, S. M., Michelson, L., and Jacob, R. (1985). Tricyclic antidepressants in obsessive–compulsive disorder: anti-obsessional or antidepressive agents? *American Journal of Psychiatry*, **142**, 572–6.

Mavissakalian, M., Jones, B., Olson, S., and Perel, J. M. (1990). The relationship of plasma clomipramine and *n*-desmethylclomipramine to response in obsessive-compulsive disorder. *American Journal of Psychiatry*, **26**, 119–22.

McDonald, R., Marks, I. M., and Blizzard, R. (1988). Quality assurance of outcome in mental health care: a model for routine care in clinical settings. *Health Trends*, **20**, 111–14.

McDougle, C. J., *et al.* (1990). Neuroleptic addition in fluvoxamine refractory OCD. *American Journal of Psychiatry*, **147**, 652–4.

Montgomery, S. A. (1980). Clomipramine in obsessional neurosis: a placebo controlled trial. *Pharmaceutical Medicine*, **1**, 189–92.

Montgomery, S. A. (1990). Is OCD diagnostically independent of both anxiety and depression? In *Obsessive compulsive disorder* (ed. S. A. Montgomery, W. K. Goodman, and N. Goeting), pp. 1–6. Duphar, Southampton.

Pato, M., Zohar-Kadouch, R., Zohar, J., and Murphy, D. L. (1984). Return of symptoms after discontinuation of clomipramine in patients with obsessive compulsive disorder. *American Journal of Psychiatry*, **145**, 1521–5.

Perse, T. L., Greist, J. A., Jefferson, J. W., Rosenfeld, R., and Dar, R. (1987). Fluvoxamine treatment of obsessive-compulsive disorder. *American Journal of Psychiatry*, **144**, 1543–8.

Prasad, A. (1974). A double-blind study of imipramine versus zimelidine in treatment of obsessive compulsive neurosis. *Pharmacopsychiatry*, **17**, 61–2.

Price, L. H., Goodman, W. K., Charney, D. S., Rasmusen, S. A., and Heninger, G. R. (1987). Treatment of severe obsessive-compulsive disorder with fluvoxamine. *American Journal of Psychiatry*, **144**, 1059–61.

Rachman, S., *et al.* (1979). The behavioural treatment of obsessional-compulsive disorders, with and without clomipramine. *Behaviour Research and Therapy*, **17**, 467–78.

Rapoport, J. (1990). *The boy who couldn't stop washing*. Collins, London.

Rasmussen, S. A. (1984). Lithium and tryptophan augmentation in clomipramine-resistant obsessive-compulsive disorder. *American Journal of Psychiatry*, **141**, 1283–5.

Robins, L. N., *et al.* (1984). Lifetime prevalence of specific disorders in three sites. *Archives of General Psychiatry*, **41**, 949–58.

Salkovskis, P. M. and Kirk, J. (1989). Obsessional disorders. In *Cognitive behaviour therapy for psychiatric problems: a practical guide* (ed. K. Hawton, P. M. Salkovskis, J. Kirk, and D. M. Clark), pp. 124–68. Oxford University Press.

Stern, R. S., Marks, I. M., Mawson, D., and Luscombe, K. K. (1980). Clomipramine and exposure for compulsive rituals. II. Plasma levels, side effects and outcome. *British Journal of Psychiatry*, **136**, 161–6.

Thoren, P., Asberg, M., Cronholm, B., Jornested, L., and Traskman, L. (1980). Clomipramine treatment of obsessive-compulsive disorder. I. A controlled clinical trial. *Archives of General Psychiatry*, **37**, 1281–5.

Turner, S. M., Jacob, R. G., Beidel, D. C., and Himmelhoch, J. (1985). Fluoxetine treatment of obsessive-compulsive disorder. *Journal of Clinical Psychopharmacology*, **5**, 207–12.

Volavka, J., Neziroglu, F., and Yaryura-Tobias, J. A. (1985). Clomipramine and imipramine in obsessive compulsive disorder. *Psychiatry Research*, **14**, 83–91.

Yaryura-Tobias, J. A. and Bhagavan, H. (1977). L-Tryptophan in obsessive-compulsive disorders. *American Journal of Psychiatry*, **134**, 1298–9.

Zohar, J., Insel, T. R., Zohar-Kadouch, R. C., Hill, J. L., and Murphy, D. L. (1988). Serotonergic responsivity in obsessive-compulsive disorder: effects of chronic clomipramine treatment. *Archives of General Psychiatry*, **45**, 167–72.

7

Anorexia nervosa: what treatments are most effective?

CHRISTOPHER P. FREEMAN and
J. RICHARD NEWTON

Introduction

Anorexia nervosa (AN) can be severe, crippling, and an often fatal disorder. Theander's follow-up study (Theander 1983) has now reached 33 years and shows an excess mortality of 22 per cent, about two-thirds dying from the direct effects of the disorder and one-third from suicide (Paton 1988). Sufferers who had had the disorder for 12 years or more rarely recovered. Anorexia nervosa is a fascinating disorder which goes against the general rules of most other psychiatric conditions. Sufferers have an increased level of educational attainment, a lower rate than average of family members with a history of psychiatric disorders (Eagles *et al.* 1990), a low rate of separation and divorce in families, and a higher prevalence in the upper social classes. No other psychiatric condition shares these demographic features, although obsessive compulsive disorder (OCD) comes closest and does share some of them.

The fascination with the disorder has led to it being of major research interest in at least five university departments of psychiatry in the UK. Many treatment programmes have been described but there have been no randomized controlled trials of acute treatment. Therapy has therefore been guided entirely by clinical opinion rather than by clinical science. It is puzzling to the authors why bulimia nervosa, a more recently described eating disorder, has had over 30 controlled trials of different types of psychotherapy and drug treatment, despite the fact that it was only de-lineated as a separate disorder following Russell's description in 1979 (Russell 1979), and that it is a less severe condition with a much lower mortality. We feel that it is time for a reappraisal of conventional treatment approaches to AN, which traditionally have been time consuming, very expensive, and remarkably ineffective. We are dealing with the most lethal of psychiatric conditions, excluding organic disorders, with a mortality higher than that of schizophrenia or manic-depressive illness. As well as the marked psychological sequelae, long-term sufferers are liable to have stunted growth, sterility, renal failure, and severe osteoporosis with subsequent fractures.

The clinical syndrome

Space precludes a detailed discussion of the clinical features which will already be well known to most readers. The DSM-III-R (APA 1987) definition summarizes the core characteristics:

1. Refusal to maintain body weight over a minimal normal weight for age and height (e.g., weight loss leading to maintenance of body weight 15 per cent below expected or failure to make expected weight gain during a period of growth leading to a body weight 15 per cent below expected).
2. Intense fear of becoming obese even when underweight.
3. Disturbance in the way in which one's body weight, size or shape is experienced, e.g., the individual claims to feel fat even when emaciated, or believes that one area of the body is too fat even when obviously underweight.
4. In females, absence of at least three consecutive menstrual cycles when otherwise expected to occur.

It is important to note that this definition stresses the psychological aspects of the disorder. The physical symptoms are all secondary to starvation and are found equally in anorexia occurring from other causes. The DSM-III-R definition changed markedly from earlier operational definitions. Weight loss has now to exceed only 15 per cent rather than 25 per cent and a duration of amenorrhoea of only three missed cycles is relatively short. This revised definition will include many adolescents who lose 8 to 10 kilograms of body weight who would never have been diagnosed as anorexic a few years ago. It is particularly important to take note of this when evaluating more recent epidemiological and treatment studies.

Case register estimates of the incidence of AN in the UK and US range from 0.37 to 4.06 per 100 000 of the population per year (Szmukler 1985). Rates in more specific populations, such as female schoolchildren or university students, suggest a prevalence of 1 to 2 per cent and the sex ratio of patients reporting to a psychiatrist with AN is in the region of 1 male for every 20 females. The peak age of onset is around 17 years and although more common in social classes I, II and III, it is seen in people from all social classes.

It is helpful to divide the psychopathology of anorexia nervosa into specific and general types. Specific psychopathology includes the markedly dysfunctional thoughts about shape and weight which lead to the 'relentless pursuit of thinness' characteristic of AN. This leads to control of weight and shape by rigorous dieting, laxative abuse, self-induced vomiting, fasting, excessive exercise, etc., leading to the marked weight loss. Appetite, however, is not lost and patients may become preoccupied with thoughts of food and eating. As weight decreases, body image disturbance increases which further drives the disorder. The pursuit of thinness is not ego-dystonic and can be seen as reasonable by the patients—indeed, anorexics characteristically judge their self-control and self-esteem purely in terms of weight. Denial of the problem and resistance to treatment is a common consequence of this.

The general psychopathology may be shared with other types of disorder. It includes many features associated with starvation (Keys *et al.* 1950), including food preoccupation, eating rituals, binges, social withdrawal, lability and depression of mood, poor concentration, irritability, stealing food, loss of energy, sleep disturbance, and obsessional symptoms. Most of the physical features of AN can be considered secondary to the starvation syndrome. William Gull first described and named anorexia nervosa in 1874 (Gull 1874) and he emphasized the importance in treatment of weight res-toration and the need to recognize the role of the family. Since then much has been written in the literature regarding aetiology, treatment, and out-come of this disorder. The Hippocratic oath states that it is our ethical duty to 'follow that system or regimen which according to our ability and judgement we consider for the benefit of our patients and to abstain from whatever is deleterious or mischievous'. Presumably, in modern medicine, this entails being guided by clinical trials but unfortunately there are few available. Our first dilemma then is where does a therapist look for guidance.

Plenty of opinions but few facts

Most of the publications describing aetiology, treatment or outcome in AN have been largely descriptive, retrospective or uncontrolled and there is no evidence that any form of intervention influences the long-term course of this disorder. In attempting to justify this statement we will first review the findings of controlled trials of psychiatric treatment and then look at outcome studies in AN.

A small number of controlled treatment trials have been reported. Russell *et al.* (1987) reported a controlled trial of either family therapy or supportive therapy in 57 patients with AN with follow-up at one year. They found that family therapy was marginally more effective than supportive psycho-therapy but only in those patients whose illness began before the age of 19 and whose duration of illness was less than three years. Patients who received brief psychotherapy were compared with a matched control group receiving only dietary advice by Hall and Crisp (1987) and at one year follow-up they performed less well in terms of weight gain. A recent study (Crisp *et al.* 1991) reported a controlled study of in-patient treatment; out-patient individual and family therapy; out-patient group therapy; and a one-off assessment with no further treatment control group. The authors concluded that all three treatment groups were highly effective at one year follow-up. It is important to highlight here, the findings from this study, that out-patient individual therapy led to the best outcome at one year follow-up of all three treatment groups, and that there was a relative failure of the in-patient regime to produce sustained weight gain. Behavioural techniques are frequently used in in-patient programmes but neither Eckert *et al.* (1979) nor Touyz *et al.* (1984) found any benefit from using strict behavioural regimes aimed at weight gain. This will be discussed later in this chapter.

The most recent major review of outcome in AN was performed by Herzog *et al.* (1988). The outstanding feature of this review was the marked variability in all outcome measures: for example, mortality ranged from 0 to 22 per cent of patients; underweightness from 15 to 43 per cent; amenorrhoea from 4 to 42 per cent; and food restriction continued in 3 to 67 per cent of patients followed-up. There was a marked lack of concordance among studies for variables predictive of outcome and no attempt was made to assess the impact of various treatments on outcome. Clearly then, AN continues to be a disorder with a high degree of chronicity and mortality. Herzog *et al.* (1988) concluded in their review that the variability in outcome of AN could be explained by the view that it is a complex multi-dimensional disorder with multiple perpetuating factors. If this is so then its treatment must also be complex and take into account these multi-dimensional factors. Garner and Garfinkel (1985) highlight the agreement on general treatment principles between the contributors despite writing about treatment regimes based on very different theoretical bases. These areas of agreement in treatment are the need for:

1. A multi-component treatment programme.
2. The reversal of starvation symptoms.
3. An open, honest, and collaborative relationship with the patients.
4. Psychotherapies to be aimed at the predisposing, precipitating, and maintaining factors of the disorder in any one individual, taking into account all of the problems that person may have.

However, within this broad basis of agreement between experts in the field there still remain many dilemmas to consider. In the foregoing discussions much of the focus is on weight restoration in very low-weight anorexics as it is in this group of patients that management dilemmas abound.

Is early treatment advantageous?

Although early intervention is thought to be desirable there is remarkably little evidence to support this and some of our most severe and long-standing cases have been in treatment from early in their anorexic career. Nevertheless, whatever model of the aetiology of AN one uses it would seem sensible to try and identify and treat cases early. When the disorder has become established for a few years it appears to become almost self-perpetuating. Many reasons may account for this. A body adapts to a starvation state. There is increasing obsessionality and intrusive thinking that accompanies starvation. As body weight decreases the image disturbance becomes more powerful. Some anorexics may become addicted to the positive feelings of starvation and have marked dysphoric feelings when they eat. As the disorder progresses behavioural and personal changes occur within the family which may reward and perpetuate the disorder. Probably different combinations of several factors apply in each case.

Early detection involves questioning about factors that tend to accompany AN but do not accompany normal adolescent dieting. These include:

1. Always wanting to eat less than others in the family, particularly female siblings and mother.
2. An intense interest in food and wanting to cook for others (most 'normal dieters' try and stay away from food).
3. Wanting to eat alone.
4. Increasing socialization in other areas.
5. Increasing perfectionism and obsessionality.
6. Increasing activity levels (compulsive exercising).
7. Hiding weight loss with big bulky clothes.
8. In public, appearing to want to minimize the amount of weight loss (most 'normal dieters' are delighted with their weight loss as is the anorexic in private).
9. Not breaking a diet at all, even for special occasions.
10. Denial of being on a diet despite obvious dieting.

Having detected a case, the appropriate interventions for a non-specialist or general practitioner might be:

1. The use of a simple questionnaire, such as the Eating Attitudes Test (EAT) (Garner and Garfinkel 1979) or Eating Disorder Inventory (EDI) (Garner *et al.* 1983), helps and the score can be reported to the individual: 'You may not think that you have AN but you scored as highly as a group of in-patients receiving in-patient treatment for AN'. (It should be noted that the EAT and EDI are not particularly good diagnostic instruments.)

2. Ask the patient to keep a food and mood diary for a week without increasing their food intake so as to get a baseline and to encourage the idea of measurement.
3. Give some brief information about the physical effects of starvation, particularly regarding fertility, osteoporosis, and growth retardation. (Many girls do not realize that they will not only be slimmer but also much shorter.)
4. Try and establish an initial agreement to stop losing weight and achieve a stable if low weight rather than at this stage the idea of weight gain.
5. Make a list of 'forbidden' foods and begin to introduce variety into the diet even if intake is not increased.
6. Give a brief explanation of some of the ideas about the aetiology of AN, e.g., fear of adulthood, fear of one's sexuality, the increasing feelings of self-control that dieting produces: 'Which one of these fits you best?'
7. Try and form a therapeutic alliance: 'Let's try and beat the AN together'; 'Let's not battle with each other'; 'Let's try and fight it (the anorexia) together rather than fight each other'.
8. Only after all these stages should one attempt to try increase food intake.
9. Introduce the ideas of specialist referral. Sometimes it is better to negotiate one-off specialist assessment which does not imply treatment. This may be more acceptable to the patient. However, you have to know your specialist and know what he or she may suggest.

What treatment setting is the most appropriate?

Most centres offering treatment for anorexia nervosa would agree that in-patient treatment is required for all anorexics whose weight loss is rapid, who are at risk of suicide, whose weight is 65 per cent below mean matched population weight for height (MMPW), and whose health is therefore at immediate risk. It is also indicated for those who have failed to respond to out-patient treatment. To most people this admission policy would intuitively appear right and have the patient's best interests at heart. These admission criteria are certainly less ethically dubious than other reasons cited for admission, such as the need to separate the patients from their families, or to take away the patient's control of eating. Clearly, admission to hospital is an important life event for any patient and should never be taken lightly.

An early review by Browning and Miller (1968) of the merits of in-patient treatment concluded that hospitalization did little to alter the long-term course of the disorder and many authors have commented on the relapse rate on discharge from the in-patient environment (e.g., Morgan *et al.*

1983). There is little doubt that in the short term in-patient treatment has immediate life-saving value for some emaciated patients.

More recently, issues of cost effectiveness have become important in the clinical environment and this has led to increasing interest in out-patient management of all patients with anorexia nervosa. From the patient's point of view the need for in-patient treatment, even for the chronic, seriously underweight patients, has been questioned by Freeman *et al.* (1990). They compared outcome in anorexics randomly allocated to in-patient or continued day-patient care. They found that in-patients gained weight more rapidly than day-patients but that this difference was lost over time. At 20 months follow-up no significant difference in general outcome measures existed between the two groups. However, it is interesting that patients who received day-patient treatment felt more in control of their eating behaviour and therefore may be less at risk of further relapse. It seems then that even severely emaciated anorexics may be treated as out-patients and that, at the very least, patients could be given a choice as to what treatment setting they prefer. Further details of this clinical programme are given below.

This brings us to another ethical problem surrounding the treatment of anorexia, namely, the issue of compulsory admission using the Mental Health Act.

Informal versus compulsory admission

If the admission guidelines mentioned above are followed there will be some patients who are resistant to treatment in an out-patient form and yet refuse voluntary admission. Should such patients be detained under the Mental Health Act? A proportion may also require to be detained involuntarily even after informally taking part in an in-patient treatment programme for some time. By necessity, much of our discussion will take place from a medico-legal standpoint. However, first of all it is appropriate to consider the problem from a philosophical base regarding the nature of anorexia nervosa. The disordered thoughts in AN regarding fear of fatness, pursuit of thinness, and body image misperceptions are the drive to the abnormal eating behaviour. It is generally agreed that these thoughts are best regarded as overvalued ideas rather than delusions, that is, an abnormal belief which is culturally acceptable and comprehensible but that is believed by the person beyond the bounds of reason. However, these thoughts are often held so intensely despite abundant evidence to the contrary that they do appear to be delusional and hence to be part of psychotic illness. This distinction is important. If AN is an illness driven by delusional ideas then most doctors would agree that sufferers are not responsible for their self-damaging behaviour [e.g., McNaughton's Rules (West and Walk 1977)].

However, psychotherapeutic interventions in AN are based on the assumption that patients have freedom of choice over their actions and in a very real sense are held answerable for them.

Most patients with AN have some understanding of the real consequences of their actions but to a large extent feel unable to prevent themselves from carrying them out. The actions of an anorexic are not caused by external forces but by internal forces and feelings of subjective compulsion. Anorexia nervosa, therefore, can be compared with drug-taking or alcohol dependence, which share similar characteristics of being self-damaging, compulsive, and relatively ego-syntonic behaviours. It is noteworthy, however, that both drug-taking and alcohol dependence are specifically excluded from definitions of mental disorders laid down in the 1983 Mental Health Act and, therefore, may be seen as slightly less worthy of treatment. For the purposes of the present discussion we have to conclude that AN is associated with only partial responsibility for patients' actions and in a pragmatic way most modern treatment regimes take this into account with emphasis both on psychotherapy and on the reversal of weight loss using engagement, education, and agreement between patient and therapist on the need for *controlled* weight gain.

If agreement cannot be made then compulsory powers may be used. In order to detain someone in hospital for one month (Section 26 in Scotland; Section 2 in England and Wales) the person has to be shown to be suffering from a mental disorder which makes it appropriate for him or her to be detained and it has to be shown that it would be in the interests of the person's own health or safety or the protection of others. In addition, for a long-term order (Section 18 in Scotland; Section 3 in England or Wales) it has to be agreed by two independent doctors that the mental disorder is of such a nature that detention is appropriate for the treatment of the mental disorder and that such treatment cannot be provided unless the patient is detained under part of the act. Anorexia nervosa clearly fits into the definition of mental disorder used in the Mental Health Act (MHA) and if the ground for detention is that of suicidal risk then few ethical problems arise. However, if the majority of detentions are based on the life-threatening complications associated with starvation then it seems reasonable to question the applicability of the MHA. Starvation and its effects are primarily physical disorders and their treatment involves no psychological expertise. Merely re-feeding the patient will remove the risk to life.

Such urgent treatment is governed by the common law duty of care owed by the doctor to his patient and from the common law doctrine of necessity. Neither require recourse to the MHA. Difficulties arise with this point of view in regard to who decides when urgent treatment is necessary and also on when to decide that the need for such urgent treatment disappears.

In practice, the length of time required to reverse the starvation status is so long that no matter which setting the treatment takes place in it would be wrong to detain the patient without providing him or her with the rights associated with the MHA. Indeed, the ward staff would rightly refuse to be involved in such detention of a patient against his or her will.

If we accept the necessity to admit some people compulsorily and yet acknowledge the impracticality of force-feeding someone against his or her will, what treatment options are available? In most circumstances, the reality presented to patients is that they will remain detained in hospital until they regain their health and this often is sufficient for them to comply 'voluntarily' with controlled re-feeding. But how ethical is this? Patients and others may view this practice simply as a form of benevolent blackmail. It certainly contradicts our earlier statements regarding an open and collaborative approach to treatment. There is much anecdotal evidence that such patients simply become determined to eat their way out of hospital and then promptly relapse. Given the evidence mentioned earlier regarding long-term outcome of hospital admission, it is possible that as therapists we are accepting the obvious short-term benefits of compulsory admission and disregarding the long-term risk to the patient inherent in such an enterprise. These risks include the breakdown of the therapeutic alliance and the distress caused by involuntary attainment of what to the patient may be an acceptably high weight. As far as long-term detention under Section 18 of the Mental Health (Scotland) Act (Section 3 in England and Wales) is concerned, the same conditions mentioned above still apply so in the long term it may be difficult to justify its repeated usage in these situations.

Choice of in-patient treatment regime

In many hospitals, an operant conditioning model of behavioural treatment is used in the in-patient management of anorexia nervosa. This model generally involves linking weight gain or 'good' eating behaviour to rewards, such as increased patient privileges. A typical regime may be strict bed rest in a bare room initially with gradual increases in visiting rights, movement about the ward, physical comforts, such as television, books or radio in the room, decreased surveillance at meal times, etc., through to full informal patient privileges and autonomy in a ward setting. If weight gain or appropriate eating behaviour does not occur then it is common for those privileges to be taken away. The mechanisms by which this is to take place are usually discussed with the patient beforehand and his or her agreement to them obtained.

This procedure is costly in terms of staff time and its effectiveness has been questioned (Bruch 1974a; Eckert *et al.* 1979; Touyz *et al.* 1984). These treatments are often seen as coercive by the patient, despite their apparent consent, and practices such as enforced bed rest or wearing night clothes have been criticized in recent years by the Mental Welfare Commission as being *de facto* detention without making recourse to the Mental Health Act (Mental Welfare Commission Report 1985). Control of themselves and their environment is an important issue for most anorexics. Bruch (1974*b*) suggested that anorexics have a 'paralysing sense of ineffectiveness' and that control of weight is a way that they may regain some sense of control over themselves. In attempts to promote weight gain and a normal eating pattern this issue has to be addressed, because in an in-patient setting patients may be overwhelmed again by a feeling of loss of control. Patients are often distressed by many of the physical manifestations of starvation, such as poor concentration, preoccupation with food, sensitivity to cold etc., and these may be used to engage patients in treatment by taking the focus away from weight and presenting treatment as a way of regaining control over these distressing symptoms. Setting of target weights for anorexics can be a very controversial matter and the cause of much dissention. By emphasizing the starvation sydrome and obtaining agreement to reverse any distressing symptoms that the patient may have, it may not be necessary to set a target weight at all. Fear of losing control over the weight leading to obesity is a common problem in treatment and again this has to be addressed sensitively. Education by a dietitian regarding nutritional requirements, agreement not to allow weight gain over a set rate (e.g., 1 kg per week) and discussion of general outcome in other patients on treatment can all help to reduce these anxieties. However, anxiety may continue, particularly during or after meals and Russell (1983) and others (e.g., Fairburn and Hope 1988) have highlighted the importance of firm, considerate nursing care at these times, and if criticized, the reliance on behavioural regimes. Anxiety management techniques may also be used in an attempt to reduce distress. The general principle of in-patient management, as in out-patient management, is to enable the patient to take control of his or her own eating behaviour and to take responsibility for maintaining a reasonable healthy weight.

Issues regarding drug treatment

In some centres, use is made of anxiolytic drugs to enable patients to continue treatment whilst regaining weight. The short-term use of benzodiazepines prior to meals is advocated for example by Bhanji (1979) and certainly this can be helpful in the first two weeks of an admission. More controversial issues arise when considering prescribing other drugs such as phenothiazines or antidepressants.

At one time the phenothiazines, particularly chlorpromazine, were advocated as an important part of any treatment plan. This was because of their anti-emetic and appetite-stimulating properties and also because of their potential for reducing the near-delusional ideas about weight and shape held by anorexics. However, several problems are associated with them. Anorexics do not have loss of appetite and so stimulating their appetite is likely to increase their concerns about loss of control over eating. Dally and Sargant (1966) observed an increase in post-discharge bulimia in anorexic patients treated with chlorpromazine. Its physical side-effects, such as hypotension and reduced fit threshold, also served to reduce enthusiasm for the use of such drugs.

It has long been recognized that anorexic patients have many symptoms of depression and that they have a higher incidence of affective disorders in their families than the general population (Piran *et al.* 1983). Some experts have tended to view anorexia nervosa as a variant of affective disorder although the study by Keys *et al.* (1950) suggested that depressive symptoms may be part of the starvation syndrome. Placebo-controlled studies of tricyclics, such as clomipramine (Crisp *et al.* 1987) and amitriptyline (Biederman *et al.* 1985), failed to show any benefits with the active drug.

Serotonergic drugs have also been considered, as serotonin plays an important part in regulating eating behaviour by a putative inhibitory action through the ventromedial nucleus of the hypothalamus (Hoebel and Leibowitz 1981). This knowledge and also the noted similarity between anorexia nervosa and obsessive compulsive disorder which does respond to serotonergic drugs (Holden 1990) have led to the use of serotonin re-uptake blocking drugs in anorexia nervosa.

Until the situation is clarified, a trial of antidepressants in general may be used in recovering anorexics who remain depressed, or in-patients at any stage of their treatment when depressive symptoms include marked suicidal ideation, and occasionally it may be worth using antidepressants for their anxiolytic activity in the initial treatment phase instead of benzodiazepines.

The chronically ill patient

A final topic which deserves mention is that of the chronically ill anorexic. Theander's follow-up study of anorexics, which has now reached its 33rd year, suggests that approximately 15 per cent remain chronically ill 12 years after onset, with few patients then recovering (Theander 1983). This subgroup of patients who are chronically starving may have had multiple, unsatisfactory (both for themselves and for staff looking after them) admissions to hospital in the past, and may be reluctant to engage in any further treatment. Despite their low weight many appear able to maintain jobs and some sort of lifestyle outside hospital. Management, therefore, should be aimed at enabling them to maintain a maximum tolerable weight, to help them deal with stresses without resorting to further weight loss and to help them accept their disability whilst still obtaining the most from their lives. This can often effectively be done in a group setting with other patients in a similar situation and we have had positive results with a long-term support group for chronic eating disorders.

Clinical guidelines

This chapter has touched on a number of difficulties that we have encountered in the management of anorexic patients. Few firm conclusions can be drawn on how to manage these problems. However, using our clinical practice as a guide we will attempt to suggest some reasonable effective ways of dealing with them.

1. *Choice of treatment setting*

Out-patient or day-patient? Almost all anorexics may be managed out of hospital. The choice of out-patient or day-patient therapy will be determined by several factors, including severity of disorder, engagement and motivation, and social circumstances (e.g., job, desirability of time away from family, etc.), and patients may move from one to the other at different times in therapy.

In-patient? This should preferably be provided in a psychiatric ward with experience of treating eating disorders and with sufficient staff and facilities to deal with medical problems. In-patient treatment can be reserved for those anorexics with marked physical complications, such as very low systolic blood pressure, low serum potassium, cardiac arrhythmias, or psychiatric complications, such as serious suicidal ideation or marked depressive symptoms. Strict behavioural regimes for in-patients are not helpful and may infringe the rights of the patient. Changeover from out- to in-patient treatment should only occur after full discussion with the patient regarding its desirability and its likely consequences and form.

Day-patient? In our treatment programme nearly all the patients are managed on an out-patient or day-patient basis. Our admission rate is approximately 2 per cent and our compulsory detention rate less than 1 per cent (between 30 and 50 per cent of anorexics are compulsorily detained in some centres). These figures do not relate to a highly selective service but to a regional catchment area with a high proportion of tertiary referrals from elsewhere in Scotland, with no competing services and, in particular, no private beds. We feel we have been able to achieve this by a collaborative approach, using no coercion, based on education, advice, and a cognitive behavioural treatment regime. We encourage an open, collaborative atmosphere by a lack of a hierarchical structure within the unit (there is no all-powerful consultant who knows best), and by open access to case notes, with encouragement of the patient to read what is said about him or her and to write comments him- or herself in the notes. We have an experimental attitude to treatment: 'Let's try, this is an experiment; if you can do it one way we will learn something, if you can do it another, we can also learn: there is no success and no failure'. Such an experiment might be trying to eat one meal as a non-anorexic or setting limits on what the patient perceives as his or her mother's intrusiveness. Everything that is done in treatment is aimed at promoting autonomy rather than reducing it and encouraging self-control rather than using coercion.

We doubt that this treatment could be carried out in an in-patient unit. The structure of such units, the nursing rules, and the demands made on nurses to behave quite differently with other groups of patients all militate against such an approach. Whether the approach would work in a general psychiatric day unit rather than a special unit remains to be tested.

2. *Use of the Mental Health Act*

In general, this should be reserved only for patients with life-threatening complications or for those patients who are at high risk of suicide and who refuse voluntary admission to hospital. Repeated admissions to hospital for weight restoration using the Mental Health Act should occur only after careful consideration of the immediate short-term benefits of admission compared with the likely long-term outcome for that patient. It may be necessary to accept that some patients will remain chronically ill and a less intensive approach that keeps these people in treatment should be adopted.

3. *Type of treatment*

Regardless of setting, all patients should take part in a multi-component treatment plan. The general philosophy of the therapeutic contract should be that of an open and collaborative endeavour in which maximum autonomy and responsibility for eating and other behaviours is accepted by the patient, enabling him or her to maintain a reasonable healthy weight and to deal with his or her life situation in a coping and adaptive manner without resorting to preoccupation with shape and weight, food restriction, etc. Included in this multi-component treatment plan should be full education regarding anorexia nervosa, nutritional and dietetic information, contact with other anorexics in a group setting, anxiety management, and some form of individual psychotherapy. The type of psychotherapy offered may depend on the practitioner's theoretical orientation and the reader is referred to the *Handbook of psychotherapy for anorexia nervosa and bulimia nervosa* (Garner and Garfinkel 1985) for further details of this. In our practice, we use a cognitive-behavioural psychotherapy which we initiate in the assessment period and often much work is done before weight gain can be achieved.

Family therapy may be reserved for patients under 18 years of age and for older patients where the family is thought to be important in maintaining the disorder. A parent and spouse support group may be a useful alternative to family therapy in that it allows parents to voice their anxiety and concerns in an educational and supportive setting without encroaching on the patient's own treatment programme.

Drug treatment Antidepressants should be used when indicated for depressive illness whilst acknowledging the major role of starvation in producing depressive symptoms in these patients. If anxiolytic medication is felt to be necessary then consideration should be given to the use of an antidepressant rather than short-term use of benzodiazepines. In general, phenothiazines should be avoided.

Other types of drug treatment should not be neglected. There is limited evidence that calcium alone protects against osteoporosis but calcium and oestrogens may have this effect. The role of bisphonate drugs has yet to be evaluated. Metoclopramide is useful in some patients. It speeds gastric emptying and reduces the feeling of being bloated that many anorexics experience after meals. There have been enthusiastic reports of the use of naltrexone, the theory being that naltrexone might abolish the endorphine-mediated high produced by starvation and the subsequent dysphoria which results after eating. Both the theory and the treatment require further evaluation.

Conclusion

Most authors would seem to agree that anorexia nervosa is a final common pathway by which many types of personal and familial distress are expressed. There does not appear to be a primary biological cause, but there may well be a biological substrate which predisposes certain individuals to be much more successful at prolonged dietary restraint. Most authors also agree that whatever the familial and personal factors involved, one step in the final common pathway is an increased feeling of self-control and mastery that comes from successful dieting. However maladaptive this may be it is nevertheless a solution for the anorexic patient. The core dilemma in treatment is how to respond with interventions that promote self-esteem, feelings of autonomy, self-control, and mastery rather than reduce them. Balancing these concerns against patients' reluctance to accept treatment, their obvious need for re-feeding, and the very high mortality and morbidity is an extremely difficult task and it is perhaps not surprising that most of us get the balance wrong.

References

APA (American Psychiatric Association) (1987). *DSM-III-R* [Diagnostic and Statistical Manual (3rd edn, revised)]. APA, Washington DC.

Bhanji, S. (1979). Anorexia nervosa: physicians' and psychiatrists' opinions and practice. *Journal of Psychosomatic Research*, **23**, 7–11.

Biederman, J., Herzog, D. B., and Rivimest, N. (1985). Amitriptyline in the treatment of anorexia nervosa: a double-blind placebo-controlled study. *Journal of Clinical Psychopharmacology*, **5**, 10–16.

Browning, C. H. and Miller, S. I. (1968). Anorexia nervosa: a study on prognosis and management. *American Journal of Psychiatry*, **124**, 1128–32.

Bruch, H. (1974a). Perils of behaviour modification in the treatment of anorexia nervosa. *Journal of the American Medical Association*, **230**, 1419–22.

Bruch, H. (1974b). *Eating disorders, anorexia, obesity and the person within*. Routledge, Kegan-Paul, London and Boston.

Crisp, A. H., Lacey, J. H., and Crutchfield, M. (1987). Clomipramine and drive in people with anorexia nervosa: an in-patient study. *British Journal of Psychiatry*, **150**, 355–8.

Crisp, A. H., et al. (1991). A controlled study of the effect of therapies aimed at adolescent and family psychopathology in anorexia nervosa. *British Journal of Psychiatry*, **159**, 325–33.

Dally, P. J. and Sargant, W. (1966). Treatment and outcome of anorexia nervosa. *British Medical Journal*, **2**, 703–5.

Eagles, J. M., Wilson, A. M., Hunter, D., and Callander, J. S. (1990). A comparison of anorexia nervosa and affective psychosis in young females. *Psychological Medicine*, **20**, 119–23.

Eckert, E. D., Goldberg, S. C., Halmi, K. A., Casper, R. C., and Davis, J. M. (1979). Behaviour therapy in anorexia nervosa. *British Journal of Psychiatry*, **134**, 55–9.

Fairburn, C. G. and Hope, R. A. (1988). Disorders of eating and weight. In *Companion to psychiatric studies* (ed. R. E. Kendell and A. K. Zealley), pp. 588–604. Churchill Livingstone, Edinburgh.

Freeman, C. P. L., Shapiro, C., Morgan, S., and Engliman, M. (1990). Anorexia nervosa: a random allocation control trial of two forms of treatment. Paper presented at Fourth International Conference on Eating Disorders, New York.

Garner, D. M. and Garfinkel, P. E. (1979). *The Eating Attitudes Test: an index of symptoms of anorexia nervosa*. Psychological Medicine, **9**, 273–9.

Garner, D. M. and Garfinkel, P. E. (1985). *Handbook of psychotherapy for anorexia nervosa and bulimia nervosa*. Guildford, New York.

Garner, D. M., Olmstead, M. P., and Policy, J. (1983). Development and validation of a multi-dimensional eating disorder inventory for anorexia nervosa and bulimia. *International Journal of Eating Disorders*, **2**, 15–34.

Gull, W. W. (1874). Anorexia nervosa. *Transactions of The Clinical Society, London*, **7**, 22–8.

Hall, A. and Crisp, A. H. (1987). Brief psychotherapy in the treatment of anorexia nervosa. *British Journal of Psychiatry*, **156**, 185–92.

Herzog, D. B., Keller, M. D., and Lavori, P. W. (1988). Outcome in anorexia nervosa and bulimia nervosa: a review of the literature. *Journal of Nervous and Mental Disease*, **176**, 131–43.

Hoebel, B. G. and Leibowitz, S. F. (1981). Brain monoamines in the modulation of self-stimulation feeding and body weight. *Research Publications. Association for Research into Nervous and Mental Diseases*, **132**, 103–42.

Holden, N. L. (1990). Is anorexia nervosa an obsessive–compulsive disorder? *British Journal of Psychiatry*, **157**, 1–5.

Keys, A., Bruzek, J., Henshel, A., Mickelson, O., and Taylor, H. L. (1950). *The biology of human starvation*. University of Minnesota Press, Minneapolis.

Mental Health (Scotland) Act (1985). *Report of the Mental Welfare Commission for Scotland*. HMSO, Edinburgh.

Morgan, H. G., Purgold, J., and Welbourne, E. J. (1983). Management and outcome in anorexia nervosa: a standardised prognostic study. *British Journal of Psychiatry*, **143**, 280–7.

Patton, G. C. (1988). Mortality in eating disorders. *Psychological Medicine*, **18**, 947–51.

Piran, N., Kennedy, S., Owens, N., and Garfinkel, P. (1983). Presence of affective disorder in patients with anorexia nervosa and bulimia nervosa. Presented at the 33rd Annual Meeting of the Canadian Psychiatric association, Ottawa.

Russell, G. F. M. (1983). Anorexia nervosa and bulimia nervosa. In *Handbook of psychiatry*, Vol. 4. *Neuroses and personality disorders* (ed. G. F. M. Russell and L. Hersov), pp. 285–98. Cambridge University Press.

Russell, G. F. M., Szmukler, G. I., Dare, C., and Eisler, I. (1987). An evaluation of family therapy in anorexia nervosa and bulimia nervosa. *Archives of General Psychiatry*, **44**, 1047–56.

Szmukler, G. I. (1985). The epidemiology of anorexia nervosa and bulimia nervosa. *Journal of Psychiatric Research*, **19**, 143–53.

Theander, S. (1983). Research on outcome and prognosis of anorexia nervosa and some results from the Swedish long-term study. *International Journal of Eating Disorders*, **2**, 167–74.

Touyz, S. W., Beaumont, P. J. V., Glaum, D., Phillips, T., and Cowie, I. (1984). A comparison of lenient and strict operant conditioning in refeeding patients with anorexia nervosa. *British Journal of Psychiatry*, **144**, 517–20.

West, D. J. and Walk, A. (eds.) (1977). *Daniel McNaughton: his trial and its aftermath*. Gaskell Books, Ashford, Kent.

8

Sex therapy: for whom is it likely to be effective?

KEITH HAWTON

Introduction

More than 20 years have passed since sex therapy was introduced with the publication of Masters and Johnson's book *Human sexual inadequacy* (Masters and Johnson 1970). This was associated with a wave of enthusiasm on both sides of the Atlantic. The positive welcome to this approach was partly due to its revolutionary nature. Thus, for the first time in the history of the treatment of sexual dysfunction an approach was available which incorporated a highly structured programme of homework assignments, counselling, and education. Moreover, both dysfunctional patients and their partners were included together in treatment, rather than individuals being treated alone as was previously the case. Enthusiasm for this approach was also stimulated by the outstanding outcome data that Masters and Johnson reported on the basis of treating more than 500 couples and individuals. Not only did they claim a remarkably low 'failure' rate at the end of therapy (18.9 per cent), but in a five-year follow up of 313 couples (all post-treatment non-failures) they found an overall relapse rate of only 5.1 per cent.

Other therapists have subsequently obtained far more modest results. This may partly reflect the degree of selection which applied to Masters and Johnson's couples. Also, the methods used by Masters and Johnson to evaluate outcome have been highly criticized (Zilbergeld and Evans 1980). Nevertheless, sex therapy has stood the test of time and is now a well-established therapeutic approach.

This chapter addresses the question of who is likely to benefit from sex therapy, a question of considerable clinical relevance in view of the time and commitment this treatment involves for both therapists and couples, the variability in the outcome, and the very real potential for harming the relationships of couples inappropriately offered this treatment. In order to examine this question one first needs to be aware of the findings regarding the outcome of sex therapy. As some readers may not be familiar with this treatment approach an outline of sex therapy is provided in the next section before outcome and prognostic factors are discussed.

The nature of sex therapy

Only a brief outline of sex therapy will be provided here (several detailed practical accounts are available, e.g., Kaplan 1974; Hawton 1985; Bancroft 1989). This chapter will be confined to couples rather than individuals with sexual dysfunction. Guidelines for treatment of individuals without partners are provided by Hawton (1985).

Before a couple are offered this treatment a careful assessment is essential, usually involving separate interviews with each of the partners alone, followed by a conjoint interview. In addition to the diagnostic purposes of the assessment, it is necessary in order to establish whether couples are suitable for this approach (see below).

The main body of sex therapy consists of: (1) *homework assignments*, of which the sensate focus programme is the core element. These are intended to help a couple gradually rebuild their sexual relationship, to improve communication, and to aid in the identification of specific factors underlying the sexual problem; (2) *psychotherapeutic work*, to help identify and modify such factors (see Hawton 1989); and (3) *education about sexuality* through didactic discussion and recommended reading (see Hawton 1985, p. 172). A certain amount of marital therapy may be combined with sex therapy when necessary, although it is not usual or advisable to conduct full programmes of sex and marital therapy simultaneously.

Treatment sessions are conducted with both partners together. Research has shown that co-therapists (one male and one female therapist), as employed in the original Masters and Johnson programme, are no more effective than single therapists (Crowe *et al.* 1981; Clement and Schmidt 1983; Mathews *et al.* 1983; LoPiccolo *et al.* 1985). Most therapists therefore work alone. Furthermore, the gender of the therapist does not appear to be important (Crowe *et al.* 1981; LoPiccolo *et al.* 1985), although most clinicians believe that there are exceptions, where, perhaps because of the specific nature of an individual partner's sexual dysfunction or personality, a therapist of one gender rather than the other is indicated (Arentewicz and Schmidt 1983). The most effective frequency of treatment sessions appears to be once or twice weekly (Clement and Schmidt 1983; Heiman and LoPiccolo 1983; Mathews *et al.* 1983), although the sessions are usually held less frequently later in the programme. Treatment of most couples is completed in 8 to 20 treatment sessions over 3 to 8 months.

The results of sex therapy

Masters and Johnson's (1970) results have already been noted. Typical post-treatment outcome figures obtained by other clinicians for couples with the usual range of sexual dysfunctions are a compliance (i.e., completion of treatment) rate of approximately two out of three couples (e.g., Hawton and Catalan 1986; Warner and Bancroft 1987; Catalan *et al.* 1990), with a positive outcome (i.e., problem resolution or marked improvement) in 50 to 60 per cent of couples who enter treatment (e.g., Duddle 1975; Bancroft and Coles 1976; Wright *et al.* 1977; Hawton and Catalan 1986; Warner and Bancroft 1987). Long-term outcome studies have revealed a subsequent positive outcome in approximately 40 to 50 per cent of couples (De Amicis *et al.* 1985; Hawton *et al.* 1986).

There are, however, marked variations in outcome for different sexual dysfunctions. Excellent results are obtained for vaginismus, with treatment gains being sustained long-term (Arentewicz and Schmidt 1983; Bramley *et al.* 1983; Hawton *et al.* 1986; Hawton and Catalan 1990), good outcome for erectile dysfunction and female orgasmic dysfunction (Arentewicz and Schmidt 1983; De Amicis *et al.* 1985; Hawton *et al.* 1986; Hawton *et al.* 1992), but less frequent and often unsustained benefits for both female and male low sexual desire, and premature ejaculation (De Amicis *et al.* 1985; Hawton *et al.* 1986; Hawton *et al.* 1991). Even though the long-term outcome for some dysfunctions may not reflect the gains which were apparent at the end of treatment, it appears that many couples experience sustained improvements in their general relationships following sex therapy (Dekker and Everaerd 1983; Hawton *et al.* 1986), possibly reflecting the major impact that this treatment may have on communication problems (Chesney *et al.* 1981; Tullman *et al.* 1981).

Thus, the results of sex therapy indicate overall satisfactory outcome in approximately 50 to 60 per cent of couples who enter treatment, but marked differences in outcome for specific sexual dysfunctions, especially in the long term. Sex therapy also often appears to benefit couples' general relationships.

The problem of deciding who should be offered sex therapy

This problem is not an insignificant one in view of the extensive commitment of time involved in sex therapy for both therapists and their patients, the variable outcome for different couples and types of sexual problems, and, not least, the danger that sex therapy, because of its highly confronting nature, might be detrimental for some couples. It is especially a problem for inexperienced therapists, who may be over-eager to offer help for people with sexual problems and who as a result may find themselves engaged in very difficult therapeutic dilemmas and disappointed by the results they achieve. This was certainly the case shortly following the publication of Masters and Johnson's results, many clinicians being misled into believing that sex therapy represented a panacea for virtually all their patients with sexual problems.

A distinction also has to be made between factors which indicate that couples are or are not suitable to enter sex therapy, and those which suggest that a positive outcome is highly likely. While, clearly, these factors overlap to some degree, there are also some differences. Although there have now been a few studies of prognostic factors in sex therapy, it has to be acknowledged that the predictive power of the factors identified is not great. Any clinician experienced in sex therapy will have treated couples for whom the prospects of success initially looked excellent but where the outcome was very disappointing, and vice versa.

Initial assessment for suitability

As noted earlier, a careful assessment must precede any offer of sex therapy (for details, see Hawton 1985, pp. 97–115). By no means all couples presenting with sexual dysfunction are suitable for this approach. Of 200 consecutive couples seen in our clinic, only 55 per cent were thought suitable, and of these almost 30 per cent declined the offer of treatment (Catalan *et al.* 1990). Some of the unsuitable couples were offered marital therapy, some brief counselling, and others treatment for psychiatric disorders.

When a couple have been referred with the expectation of receiving sex therapy but are clearly unsuitable the reasons need to be explained frankly yet sympathetically. General practitioners need to be careful whom they refer, and are advised to discuss over the telephone referrals about whom they are uncertain before committing couples to an assessment, which in itself can sometimes be a fairly uncomfortable experience especially if it does not result in treatment.

The following are specific factors which most clinicians would agree determine suitability or unsuitability for sex therapy.

Clearly defined sexual problems and goals　　The therapist and both partners must be able to agree on the nature of the sexual problem to be addressed during treatment, although it is neither necessary nor often possible at the outset to have a full understanding of the factors which have caused it or are maintaining it. Vagueness about the nature of a sexual problem may indicate inadequacy of the therapist's assessment or lack of awareness of one or both partners that the problem is not primarily sexual, but reflects some other difficulty, such as general relationship issues. Some couples find it easier to present with an apparent sexual problem rather than admit that there are fundamental difficulties in their overall relationship. It is also important that the goals of therapy are reasonable (e.g., it is unlikely that a woman with a long history of low sexual desire will become ever-eager to have sex with her partner, or that a man with premature ejaculation will never again experience rapid ejaculation).

Reasonable general relationship It would be naïve to expect that all couples with treatable sexual dysfunctions should have excellent general relationships. Apart from other considerations, prolonged sexual diffi- culties often result in more general problems. However, couples whose sexual dysfunctions are the *result* of major general relationship problems are highly unlikely to benefit from sex therapy and may indeed, because of its highly confronting nature, be further traumatized by it.

If either partner is having a current affair this certainly precludes sex therapy for the time being, although treatment might be appropriate if the affair is terminated.

A general rule of thumb is that a couple's relationship should be suf- ficiently harmonious that there are reasonable prospects of the partners being able to engage collaboratively in the treatment programme, especially the initial homework assignments. Where the therapist is in doubt about this an agreed brief initial trial period (three to five sessions) is advisable.

Reasonable motivation If it is obvious that the motivation of either or both partners is poor, sex therapy should not be offered, although it might be appropriate at a later stage when, for example, the relationship has improved as a result of marital therapy. However, a reasonable degree of healthy scepticism is to be expected in many cases and should not preclude treatment. Often couples or individual partners cannot initially see how sex therapy will help or do not really understand how it might work. Again, an agreement to give this approach a trial for a few sessions can be a helpful way of facilitating the initial phase of treatment, the purposes and benefits of sex therapy frequently then becoming more apparent.

Cultural factors are often important in determining whether couples find sex therapy acceptable. Many Asian and Afro-Caribbean males are not prepared to accept psychological treatment, and may demand a physical remedy. They may also not be prepared to involve their partners in therapy. Language barriers may further obstruct treatment. Under such cir- cumstances it can be very helpful to have access to therapists familiar with such couples' ethno-cultural backgrounds who may have more success in engaging them in appropriate treatment.

Psychiatric disorder Sex therapy is inappropriate if either partner is currently suffering from a severe acute psychiatric disorder. However, some degree of depression or anxiety commonly accompanies sexual dysfunc- tions, and may have an aetiological role, especially in low sexual desire (Schreiner-Engel and Schiavi 1986). Mild to moderate affective symptoms need not preclude sex therapy. However, where specific treatment for an affective disorder is indicated (e.g., with antidepressants), it is wise to delay sex therapy until the symptoms have improved.

Serious alcohol abuse by either partner should definitely preclude sex therapy, although the situation might be reassessed once reasonable control or abstinence are established.

Physical illness Sex therapy is often appropriate where one partner has a chronic physical disorder, even if this has contributed to the sexual problem (e.g., diabetes, multiple sclerosis). However, during the assessment of couples it is very important to look for physical disorders. Many men with erectile dysfunction, for example, have a physical condition (e.g., Spark *et al.* 1980; Catalan *et al.* 1990). Sex therapy should not begin while a condition possibly linked to the sexual problem is under investigation. This is partly because a full diagnosis may mean that another treatment approach is indicated, or that it is necessary to adjust the goals of therapy to take account of the limitations (on sexual response, for example) imposed by the physical condition.

Pregnancy It is unwise to begin sex therapy if the female partner is pregnant, because of the substantial effects pregnancy has on sexual interest in many women, especially during the third trimester. A sensible approach is to reassess the couple six months or so after childbirth in order to determine whether therapy is still needed.

Factors associated with the outcome of sex therapy

Once couples have entered sex therapy, there are then the questions of, first, who is likely to comply with and complete the programme, and secondly, who will actually benefit from it.

Compliance

Clearly, factors associated with compliance overlap those associated with positive outcome. However, there are some specific factors associated with compliance (Hawton and Catalan 1986; Catalan *et al.* 1990). One is *social class*, more couples of lower socioeconomic status tending to drop out of therapy than those of higher status. This finding, which appears to apply particularly to men with erectile dysfunction (Hawton *et al.* 1992), may reflect dissonance between the model of sexuality embodied in sex therapy and that held by many in this group of patients. It may also indicate that other approaches should be pursued (e.g., physical treatment for erectile dysfunction). Further factors associated with compliance include the *male partner's pre-treatment motivation*, the degree of *anxiety experienced by the presenting partner*, the *quality of the general relationship*, and *early progress in treatment* (Hawton and Catalan 1986; Catalan *et al.* 1990).

Factors associated with success or failure of therapy

Several factors have been identified as having prognostic significance in relation to outcome following sex therapy in couples with a range of sexual dysfunctions. The most consistent finding is that outcome is strongly associated with the quality of couples' *general relationships* (Lansky and Davenport 1975; Fordney-Settlage 1975; Mathews *et al.* 1976; O'Conner 1976; Snyder and Berg 1983; Hawton and Catalan 1986). This emphasizes the need to be careful in the initial assessment of a couple's general relationship. Interestingly, Hawton and Catalan (1986) found that it was only the female partners' ratings of the relationship which were predictive. As there were no marked differences between the mean general relationship ratings by the male and female partners in their series of couples, it appears that the women may have made more accurate ratings. This suggests that in assessing couples, particular heed should be paid to the female partners' views of the general relationship. The importance of the general relationship for the success of sex therapy raises the question of whether some couples might benefit from an initial phase of marital therapy before sex therapy. In a study of couples with disturbed general relationships in addition to sexual dysfunction, Zimmer (1987) produced evidence to support this suggestion.

Therapists' pre-treatment ratings of *motivation* are indicators of outcome, although the *male partner's motivation* appears to be the main factor (Hawton and Catalan 1986). It is possible that the model of sexuality embodied in sex therapy is more acceptable to women than men, with the result that the male partner's motivation is a major determinant of whether a couple engage effectively in the treatment programme. Therapists should be sensitive to gender differences such as this and be prepared to be flexible in their treatment approach.

The *overall quality of a couple's sexual relationship* also appears to be predictive of outcome, although not strongly so (Hawton and Catalan 1986). More specifically, Whitehead and Mathews (1977) found that pre-treatment ratings by partners of how *attractive* they found each other were also associated with outcome.

Some studies have suggested that current *psychopathology* is often associated with poor initial outcome of sex therapy (Meyer *et al.* 1975; O'Connor 1976; Levay and Kagle 1977). Hawton and Catalan (1986) were unable to replicate to this finding, in spite of employing both clinical interview and questionnaire measures, this possibly being the result of very careful selection of couples. However, in their study, poor long-term outcome was associated with either partner having a positive psychiatric history before sex therapy, this finding largely reflecting a positive history in the female partners.

Early progress in therapy, in terms of engagement in the homework assignments by the third treatment session, often provides an indication of eventual outcome (Hawton and Catalan 1986). This supports the recommended clinical practice of planning an early 'review session' in the treatment of all couples, it being agreed beforehand that this is a time when either the therapist or the couple should feel free to terminate therapy if progress has not been made and it appears that the approach is unlikely to succeed (Hawton 1985, p. 126).

Inconsistent findings have been reported with regard to the importance for outcome of the *duration of the sexual problem* (Lansky and Davenport 1975; O'Connor 1976; Hawton and Catalan 1986). Certainly it does not appear to be a major factor.

It is important to note that although the above factors are statistically associated with outcome, they are not, either alone or in combination, totally predictive of success or failure. They should therefore only be used to guide the clinician in assessing suitability for treatment and when trying to decide if a couple already in treatment are likely to benefit from continuing with it.

It is also important to note that outcome of treatment does not appear to be associated with the ages of partners, whether one or both have a sexual problem, how well informed the partners are about sexuality, and presence or strength of religious beliefs (Hawton and Catalan 1986).

Factors associated with outcome of specific sexual dysfunctions

Although the prognostic factors discussed above are the most important, some further factors have been identified with regard to the outcome of individual dysfunctions.

Erectile dysfunction   In addition to some of the prognostic factors already mentioned, Hawton *et al.* (1992) found that outcome of erectile dysfunction was particularly related to the female partner's interest in, and enjoyment of, sex prior to treatment. Thus the female partner's sexual adjustment may be at least as relevant to outcome as that of the male partner. Masters and Johnson (1970) found that the uncommon condition of primary erectile dysfunction (i.e., erectile difficulty which has been present since first sexual activity) often responds poorly to therapy. This may well be due to this problem commonly having an organic basis.

Low sexual desire   In a study of 60 couples who entered sex therapy because of the female partners' low sexual desire (Hawton *et al.* 1991), poorer outcome was associated with younger age of the male partner, shorter duration of the sexual problem, a history of the couple having separated in the past, and poor communication by the male partner. Whitehead and Mathews (1986) found that degree of attraction between partners, the quality of the general relationship, and the extent to which the female partners reported experiencing pleasant and unpleasant feelings in various sexual situations were predictors of outcome in 'sexually unresponsive' women.

Thus, factors reflecting the quality and stability of the relationship appear to be particularly important in the outcome of low sexual desire.

Vaginismus   The outcome for vaginismus is usually so good that no major pre-treatment indicators of outcome are found (Hawton and Catalan 1990). The excellent outcome probably partly reflects the generally positive characteristics of couples with this problem in terms of their general relationships, sexual adjustment other than the specific sexual problem, psychological state, and motivation (Duddle 1977; Catalan *et al.* 1990; Hawton and Catalan 1990). It is also likely to be related to the nature of vaginismus, which usually appears to be a specific phobia. Clinical impression suggests that where vaginismus is part of a more global impairment of the female partner's sexual adjustment then couples tend to respond less well to sex therapy. This may be due to women in this category tending often to have had very repressive or traumatic sexual backgrounds.

Orgasmic dysfunction   On the basis of studies of women with orgasmic dysfunction who were treated with group therapy it appears that outcome is better in those of relatively young age, where the problem is primary, and where the relationship with the partner is of relatively short duration (Schneidman and Maguire 1976; Barbach and Flaherty 1980). Conversely, outcome is likely to be poorer if the problem is either very entrenched or has developed during the present relationship, the latter probably reflecting other relationship difficulties.

Clinical guidelines

1. Sex therapy is clearly not a universal panacea for all couples with sexual dysfunction. It is essential that a careful assessment be conducted before this treatment is offered. Inappropriate treatment may encourage unreasonable expectations in couples and may sometimes be damaging for them. It is also wasteful of therapeutic time and is likely to cause therapists to become disillusioned with what is otherwise clearly a valuable treatment approach.

2. The following factors usually indicate that sex therapy should not be offered, at least until they have resolved: marked general relationship

difficulties (especially if these represent the fundamental problem); either partner having a current affair; clear absence of motivation; current major psychiatric disorder (including alcohol abuse); unclarified physical disorder; and pregnancy. The nature of a couple's sexual problem and goals should also be clearly established before sex therapy is suggested.

3. In assessing whether couples are likely to comply with treatment, attention should be paid particularly to their general relationships and their motivation, especially of the male partner. For couples of lower socioeconomic status who clearly find the approach not to their liking the therapist should consider another approach (e.g., brief counselling, physical treatment).

4. When uncertain about whether couples are suited to sex therapy the therapist might explicitly suggest a brief trial of treatment. For couples with some degree of general relationship difficulties but for whom sex therapy also appears to be indicated, a brief period of marital therapy should precede sex therapy.

5. Treatment studies have shown that once a couple enter sex therapy, outcome is particularly likely to be associated with the quality of their general relationship (especially as assessed by the female partner), the male partner's motivation and their degree of early engagement in the treatment programme. Further factors have been identified for specific dysfunctions. The known prognostic factors can assist clinicians in deciding whether or not to start treatment and in identifying aspects of couples which require special therapeutic attention. They can also help when trying to decide whether or not to continue with treatment when it has not been progressing well.

Acknowledgements

This chapter was prepared while the author was Visiting Boerhaave Professor in the Department of Psychiatry, Leiden University, The Netherlands. He thanks the University for its support and is also grateful to Joan Kirk and Michiel Hengeveld for their helpful comments.

References

Arentewicz, G. and Schmidt, G. (ed.) (1983). *The treatment of sexual disorders*. Basic Books, New York.

Bancroft, J. (1989). *Human sexuality and its problems* (2nd edn). Churchill Livingstone, Edinburgh.

Bancroft, J. and Coles, L. (1976). Three years' experience in a sexual problems clinic. *British Medical Journal*, i, 1575–7.

Barbach, L. G. and Flaherty, M. (1980). Group treatment of situationally orgasmic women. *Journal of Sex and Marital Therapy*, **6**, 19–29.

Bramley, H. M., Brown, J., Draper, K. C., and Kilvington, J. (1983). Non-consummation of marriage treated by members of the Institute of Psychosexual Medicine: a prospective study. *British Journal of Obstetrics and Gynaecology*, **90**, 908–13.

Catalan, J., Hawton, K., and Day, A. (1990). Couples referred to a sexual dysfunction clinic: psychological and physical morbidity. *British Journal of Psychiatry*, **156**, 61–7.

Chesney, A. P., Blakeney, P. E., Chan, F. A., and Coley, C. M. (1981). The impact of sex therapy on sexual behaviors and marital communication. *Journal of Sex and Marital Therapy*, **7**, 70–9.

Clement, U. and Schmidt, G. (1983). The outcome of couple therapy for sexual dysfunctions using three different formats. *Journal of Sex and Marital Therapy*, **9**, 67–78.

Crowe, M. J., Gillan, P., and Golombok, S. (1981). Form and content in the conjoint treatment of sexual dysfunction: a controlled study. *Behaviour Research and Therapy*, **19**, 47–54.

De Amicis, L. A., Goldberg, D. C., LoPiccolo, J., Friedman, J., and Davies, L. (1985). Clinical follow-up of couples treated for sexual dysfunction. *Archives of Sexual Behavior*, **14**, 467–89.

Dekker, J. and Everaerd, W. (1983). A long-term follow-up study of couples treated for sexual dysfunctions. *Journal of Sex and Marital Therapy*, **9**, 99–113.

Duddle, C. M. (1975). The treatment of marital psycho-sexual problems. *British Journal of Psychiatry*, **127**, 169–70.

Duddle, C. M. (1977). Etiological factors in the unconsummated marriage. *Journal of Psychosomatic Research*, **21**, 157–60.

Fordney-Settlage, D. S. (1975). Heterosexual dysfunction: evaluation of treatment procedures. *Archives of Sexual Behavior*, **4**, 367–87.

Hawton, K. (1985). *Sex therapy: a practical guide*. Oxford University Press.

Hawton, K. (1989). Sexual dysfunctions. In *Cognitive behaviour therapy for psychiatric problems: a practical guide* (ed. K. Hawton, P. M. Salkvoskis, J. Kirk, and D. M. Clark), pp. 370–405. Oxford University Press.

Hawton, K. and Catalan, J. (1986). Prognostic factors in sex therapy. *Behaviour Research and Therapy*, **24**, 377–85.

Hawton, K. and Catalan, J. (1990). Sex therapy for vaginismus: characteristics of couples and treatment outcome. *Sexual and Marital Therapy*, **5**, 39–48.

Hawton, K., Catalan, J., Martin, P., and Fagg, J. (1986). Long-term outcome of sex therapy. *Behaviour Research and Therapy*, **24**, 665–75.

Hawton, K., Catalan, J., and Fagg, J. (1992). Sex therapy for erectile dysfunction: characteristics of couples, treatment outcome and prognostic factors. *Archives of Sexual Behavior*, **21**, 161–75.

Hawton, K., Catalan, J., and Fagg, J. (1991). Low sexual desire: sex therapy results and prognostic factors. *Behaviour Research and Therapy*, **29**, 217–24.

Heiman, J. R. and LoPiccolo, J. (1983). Clinical outcome of sex therapy: effects of daily v. weekly treatment. *Archives of General Psychiatry*, **40**, 443–9.

Kaplan, H. S. (1974). *The new sex therapy*. Bailliere Tindall, London.

Lansky, M. R. and Davenport, A. E. (1975). Difficulties in brief conjoint treatment of sexual dysfunction. *American Journal of Psychiatry*, **132**, 177–9.

Levy, A. N. and Kagel, A. (1977). A study of treatment needs following sex therapy. *American Journal of Psychiatry*, **134**, 970–3.

LoPiccolo, J., Heiman, J. R., Hogan, D. R., and Roberts, C. W. (1985). Effectiveness of single therapists versus co-therapy teams in sex therapy. *Journal of Consulting and Clinical Psychology*, **53**, 287–94.

Masters, W. H. and Johnson, V. E. (1970). *Human sexual inadequacy*. Little, Brown, Boston.

Mathews, A. *et al.* (1976). The behavioural treatment of sexual inadequacy: a comparative study. *Behaviour Research and Therapy*, **14**, 427–36.

Mathews, A., Whitehead, A., and Kellett, J. (1983). Psychological and hormonal factors in the treatment of female sexual dysfunction. *Psychological Medicine*, **14**, 83–92.

Meyer, J. K., Schmidt, C. W., Lucas, M. J., and Smith, E. (1975). Short-term treatment of sexual problems: interim report. *American Journal of Psychiatry*, **132**, 172–6.

O'Connor, J. F. (1976). Sexual problems, therapy and prognostic factors. In *Clinical management of sexual disorders* (ed. J. K. Meyer), pp. 74–98. Williams and Wilkins, Baltimore.

Schneidman, B. and Maguire, L. (1976). Group therapy for non-orgasmic women: two age levels. *Archives of Sexual Behavior*, **5**, 239–47.

Schreiner-Engel, P. and Schiavi, R. C. (1986). Life-time psychopathology in individuals with low sexual desire. *Journal of Nervous and Mental Disease*, **174**, 646–51.

Snyder, D. C. and Berg, P. (1983). Predicting couples' response to brief directive sex therapy. *Journal of Sex and Marital Therapy*, **9**, 114–20.

Spark, R. F., White, R. A., and Connolly, P. B. (1980). Impotence is not always psychogenic. *Journal of the American Medical Association*, **243**, 750–5.

Tullman, G. M., Gilner, F. H., Kolodny, R. C., Dornbush, R. L., and Tullman, G. D. (1981). The pre- and post-therapy measurement of communication skills of couples undergoing sex therapy at the Masters and Johnson Institute. *Archives of Sexual Behavior*, **10**, 95–109.

Warner, P. and Bancroft, J. (1987). A regional clinical service for psychosexual problems: a three-year study. *Sexual and Marital Therapy*, **2**, 115–26.

Whitehead, A. and Mathews, A. (1977). Attitude change during behavioural treatment of sexual inadequacy. *British Journal of Social and Clinical Psychology*, **16**, 275–81.

Whitehead, A. and Mathews, A. (1986). Factors related to successful outcome in the treatment of sexually unresponsive women. *Psychological Medicine*, **16**, 373–8.

Wright, J., Perreault, R., and Mathieu, M. (1977). The treatment of sexual dysfunction. *Archives of General Psychiatry*, **34**, 881–90.

Zilbergeld, B. and Evans, M. (1980). The inadequacy of Masters and Johnson. *Psychology Today*, *August*, 29–43.

Zimmer, D. (1987). Does marital therapy enhance the effectiveness of treatment for sexual dysfunction? *Journal of Sex and Marital Therapy*, **13**, 193–209.

9

Patients with chronic somatization: what can the psychiatrist offer?

CHRISTOPHER BASS

Introduction

Somatization has been defined as 'the expression of personal and social distress in an idiom of bodily complaints with medical help-seeking' (Kleinman and Kleinman1985). Most patients who somatize also experience symptoms of anxiety and depression, but do not complain about these symptoms unless the doctor asks about them.

Somatization is best seen as a *process* rather than a disease entity, and it can be acute, subacute, or chronic (for the purposes of this chapter these periods are less than 4 weeks, 1–24 months, and in excess of 2 years, respectively: Rosen *et al.* 1982).

Whereas the vast majority of patients with acute and subacute somatization will have either adjustment disorders or affective disorders (anxiety with or without panic, and depression), those with chronic somatization are more likely to have either chronic relapsing/remitting affective disorders that present somatically, or one of the somatoform disorders (American Psychiatric Association 1987). They are also more likely to have a coexisting personality disorder.

Relationship between acute and chronic somatization

There is no generally accepted threshold which separates acute or subacute from chronic forms of somatization: this issue is hardly mentioned in published work. Pain research tends to use the arbitrary period of six months to define chronic pain, and this may be appropriate for somatization. However, the period of two years will be used in this chapter.

The factors that determine the chronicity in somatization have attracted little interest, but are likely to be a consequence of factors both in the patient and in the doctor. A wide range of patient factors are likely to be involved, including:

(1) the nature and cause of the somatic complaint(s);
(2) illness beliefs and attitudes, for example, strong disease conviction and an unwillingness to accept appropriate reassurance and explanation are likely to predict a chronic course; and
(3) predicamental factors, which include options available to the patient to resolve his or her difficulties.

Patients trapped in hopeless relationships, adverse social circumstances or compensation claims are more likely to have a worse outcome.

Factors in the doctor also contribute, and it is important to recognize this interactive component in chronic somatization. Inappropriate and ill-timed reassurance and advice, the prescription of unnecessary medication, and the ordering of excessive investigations are all likely to perpetuate the process.

Epidemiology

There is lack of precise information about the prevalence of chronic somatization. This is a consequence of the uncertainty about the definition and the threshold of chronicity, and the difficulty in specifying operational criteria for processes such as somatization. The following must be taken as very rough estimates.

Community-based studies

The Epidemiological Catchment Area studies in the US are based on total adult population samples which are not aimed at assessing chronic somatization as such. However, they do include findings for somatization disorder (by definition a chronic condition) with prevalence estimated at between 0.38 and 4.4 per cent depending on criteria for 'caseness' (Swartz *et al.* 1986; Escobar *et al.* 1987). These figures almost certainly represent a minority of the total chronic somatizers in the population.

Primary care settings

It has been estimated that between 26 and 30 per cent of general practice attenders are somatizers, with only minor differences between ethnic groups (Bridges and Goldberg 1985; Bhatt *et al.* 1989).

Hospital studies

Recent studies of general hospital patients referred to liaison psychiatry services report that about 40 per cent are somatizers and 15 per cent have somatoform disorders (de Leon *et al.* 1987). Somatoform disorders have been diagnosed in 12 to 53 per cent of patients attending pain clinics (Benjamin *et al.* 1988).

What can the psychiatrist offer?

This section on management will *not* be concerned with acute and subacute somatizers in primary care. A treatment 'package' for such patients has been developed by Goldberg and his colleagues, and has been described elsewhere (Goldberg *et al*. 1989; Gask *et al*. 1989). Instead, the treatment of chronic somatizers will be described, first in primary care and then in tertiary care settings.

Treatment of chronic somatizers in primary care

There has been relatively little controlled research testing the effectiveness of psychiatric interventions with chronic somatizers in primary care. This is regrettable, because such research presents opportunities for psychiatrists to demonstrate the potential cost effectiveness of their interventions. There are two important studies, both from the US, one with impressive and the other with less sanguine findings.

In the first, Smith *et al*. (1986) identified patients meeting DSM III criteria for somatization disorder or SD (APA 1980). These patients with SD have multiple, medically unexplained physical symptoms which begin in early adult life. The authors forwarded a psychiatric consultation report to the patient's primary care physicians on a randomized basis, explained the diagnosis, and suggested giving the patients fixed-interval appointments to alter symptom-contingent consulting patterns. Physicians were also advised to reduce medication and not to order laboratory tests or hospitalization unless there was a clear indication. Provision of a consultation report, in the form of a letter, was associated with reduced health care utilization and costs without worsening of the patient's clinical state or satisfaction with care.

The results of this study were impressive but difficult to explain. The intervention was little more than a letter containing sound advice, yet it led to considerable financial savings. One possible explanation is that the intervention led to significant changes in the behaviour of the *doctors* rather than the somatizing patients. Indeed, it is possible that these fee-for-service physicians were high utilizers of secondary medical care at baseline but, as a consequence of the intervention, became more conservative and reduced unwarranted tests and need for in-patient care.

The second study was in many ways more ambitious and comprehensive, yet led to *less* impact on medical utilization. In this study Katon and his colleagues (1991) randomized 251 'distressed high utilizers' of two primary care clinics to receive either a psychiatric intervention or usual care.

Each patient in the intervention group received a 30 minute interview conducted by the psychiatrist *with the primary care physician present.*

Salient points in the social and developmental history were reviewed and both physicians jointly formulated a treatment plan and negotiated a mutually acceptable course of action with the patient. This might include: initiating treatment with tricyclic antidepressants; referral to the mental health service; and arranging regular visits with the patient rather than allowing symptom-contingent visits. All primary care physicians were provided with articles on treatment of specific psychiatric disorders and were visited once during the course of the study to review each patient's progress.

At follow-up 13 months after randomization the only significant difference between the intervention and control group was in the number of primary care physician prescriptions for antidepressants (and compliance with this treatment) compared with the 18 month period before the study. There was no significant difference between intervention patients and controls at 6 and 12 months after randomization in measures of psychiatric distress, functional disability, or in any measure of utilization (visits to the surgery, use of radiographic and laboratory services, admissions to in-patient medical care).

These findings are difficult to explain, because the intervention was in many ways more comprehensive and systematic than that carried out by Smith *et al.* (1986). Possible explanations include: (1) high rates of coexisting physical disorders (two-thirds of the patients had one or more chronic physical illnesses); (2) high rates of chronic and severe psychiatric illnesses; and (3) conservative practice patterns of the primary care physicians compared with the fee-for-service physicians in Smith's study. In Katon's study, depression was usually a relapsing-remitting illness with persistent 'background' features, such as insomnia, fatigue, and chronic pain, that remained after the acute episode had subsided.

Another possible explanation for these negative findings was that *behavioural factors* in the patients received insufficient attention. Although the intervention resulted in impressive changes in the prescribing habits of doctors, those factors *maintaining* complaint behaviour and high use of health care in patients were unaltered. These factors may require more detailed assessment in future studies.

The treatment of chronic somatization in psychiatric practice

In this section I will attempt to show that systematic intervention studies involving relatively homogeneous groups of patients with chronic somatization have been effective. Indeed, there is some evidence that the more homogeneous the group, the more effective the intervention.

However, there are some key issues in the treatment of patients with chronic somatization that need to be addressed before these studies are discussed. In particular, management will not be effective if the patient is not satisfactorily engaged in treatment. For this to occur a *therapeutic alliance* needs to be established. This is the most crucial part of management, and requires considerable clinical skill. As many of these patients have medical diseases and personality disorders in addition to psychiatric illness, a multi-axial assessment is recommended (Mezzich 1988). In this section I shall describe some of the techniques we have used at King's College Hospital (KCH) during the past five years to *assess* and *engage* patients with chronic somatization (in practice the two activities overlap).

Introducing and facilitating the assessment   An important initial task is to discover the patient's attitude to the referral. Most patients resent the implication of mental illness implicit in referral to a psychiatrist and may forcefully express this antipathy (House 1990). A helpful way to elicit the patient's co-operation is to explain: 'Dr X has suggested that it might be worthwhile exploring whether emotional factors are contributing to your problems. We often find that stress can aggravate headaches, chest pain, etc., and so a psychological assessment might help you to understand your complaints better. Then we could discuss whether any treatment might be helpful'.

At the initial interview it may also be necessary to discuss with the patient his or her previous, often unsatisfactory, medical experiences in a non-critical fashion. During this phase of the interview there may be opportunities to empathize with the patient, whose fruitless rounds of multiple out-patient clinics will invariably have led to frustration and disillusionment with the medical profession. It is important, however, not to get involved in criticizing medical colleagues.

General assessment   This requires some *flexibility* on the part of the interviewer, and a grasp of the requisite interviewing skills is essential (Bird and Cohen-Cole 1990). These include *empathic witnessing*; being able to *guide and control the interview* if the patient becomes excessively circumstantial; repeatedly *summarizing* and *reflecting back* when the patient discloses information about somatic complaints that might be linked to life events (see later); and recognizing and reflecting the patient's feeling *as soon as it appears*. It is important not to stick too rigidly to a conventional psychiatric 'history-taking scheme' in these patients. As most interviews are time-limited, more than one interview is invariably needed to make a full assessment.

At KCH we use two research instruments to facilitate this process of engagement. The first is the Diagnostic Interview Schedule or DIS (Robins *et al.* 1981), which includes a list of 35 somatic complaints (in the somatoform section). The interviewer systematically enquires after each of these and asks if the patient ever experienced, for example, abdominal pain, records when it occurred and whether the patient ever visited a doctor, underwent investigations or received medication, and the impact of the illness on the patient's life. We also routinely ask whether the somatic complaints were accompanied by any symptoms of emotional distress. Whenever appropriate, an attempt should also be made to *link the physical and psychological symptoms to relevant life events* or setbacks, using judiciously timed summarizing statements inviting comment from the patient. For example, it might be appropriate to say: 'You told me that the chest tightness, tiredness, and headaches began about three years ago soon after you lost your job, and at about that time you were also feeling very tense and tearful. Have I got that right?' This approach should help the patient to recognize that: (1) each physical complaint is being taken seriously; (2) physical and emotional symptoms sometimes coexist; and (3) both might be related temporally to a specific psychosocial stressor (in this case loss of job). This latter technique, called 'making the link', encourages the patient to establish links between his or her emotional state and somatic symptoms (Goldberg *et al.* 1989).

We also *routinely* acquire the family doctor's notes as well as the medical records (often from more than one hospital), to check whether the patient's disclosures about somatic complaints are related to relevant organic pathology. Despite its many shortcomings as a research instrument, the DIS allows the interviewer to establish a *developmental perspective* and to 'quantify' the somatic complaints reported during the patient's lifetime.

It is important to negotiate this phase of the interview (which may take anything between 2 and 6 hours), before moving on to a systematic exploration of psychosocial issues. We use a semi-structured questionnaire for this purpose, which covers details of childhood history, with particular emphasis on childhood experience of illness and its impact on schooling and social development. Following this, any history of illness in the patient's mother or father should be enquired after, with particular emphasis on each parent's ability to work or fulfil a parental role. We routinely ask each patient to provide specific *examples* of the ways in which illness in a parent led to impairment in his or her working capacity and ability to care for the child (fulfil parental role).

This leads naturally to an assessment of the *quality of care and protection* received from each parent during the patient's formative years (up to 16 years), which may be facilitated by use of the Parental Bonding Instrument or PBI (Parker *et al.* 1979). This 25-item questionnaire is completed by the patient *in the presence of the interviewer*, and it provides many opportunities to discuss early childhood memories. In our experience this questionnaire often elicits affect-laden disclosures by patients, which provide an opportunity to make empathic statements. These should help to establish the therapeutic alliance. More systematic questioning about physical abuse and neglect, child sexual abuse, and contact with child psychiatric services should follow.

Because these patients are reluctant to discuss psychosocial issues we leave a detailed assessment of the past psychiatric history to the end of the interview. Previous episodes of psychiatric illness should be enquired after, as should any forensic history or parasuicide. The past medical history will have been covered by the DIS, and may reveal unnecessary surgery, in particular hysterectomy in women, as well as evidence of iatrogenic illness. Chronic somatizers habitually consume large amounts of medication, especially analgesics, and it is not uncommon for some to be in receipt of a drug for which there is no indication (e.g., anti-anginal medication for non-cardiac chest pain).

Next, details of the current illness episode need to be elicited. *Illness beliefs and attitudes* can be explored using the 14-item Whiteley Index (Pilowsky 1967). This provides further opportunities to explore general beliefs about the nature and meaning of symptoms (as well as attitudes to the medical profession, which may have become very negative as a consequence of repeated normal investigations).

Behavioural assessment It is important to make a detailed assessment of *behaviours* which are the consequence of patients' symptoms or anxiety, especially if cognitive-behavioural treatment is being planned (Salkovskis 1989). This includes what patients actually do (e.g., lie down for 16 hours each day), but also other less obvious voluntary actions (e.g., distraction, reading medical textbooks). *Reassurance seeking* from medical and non-medical sources should also be assessed, and any behavioural avoidance noted (e.g., avoidance of all forms of exercise).

A period of self-monitoring may also be helpful, especially in patients with chronic fatigue states or chest pain syndromes that are associated with considerable disability. When self-monitoring is begun the patient is asked to keep records about the relevant variables (e.g., the target problem, such as the severity of fatigue, thoughts associated with episodes, general mood, and behaviour)—see Kirk (1989) for further details. Medication use should be recorded in self-monitoring, and can be regarded as an illness behaviour which fosters preoccupation, sometimes because of unwanted side-effects.

Engagement in treatment By the time this comprehensive assessment has been carried out, the therapist should have developed a fairly clear idea of the patient's problems and should be able to *present a summary of what the patient has disclosed*, always inviting comment. Patient and therapist should come to some kind of agreement about this summary/formulation before treatment can proceed. Next, the patient and therapist must *agree on treatment goals*. During this phase different expectations of treatment and how it should proceed should be reconciled, otherwise treatment will not be effective. Finally, the patient must agree to have *no further tests or investigations*, and appreciate that during the course of treatment further lengthy discussions of symptoms will be unproductive.

What type of treatment?   The type of treatment depends on those factors that are *maintaining* complaint behaviour, which may range from marital or family discord; chronic untreated depressive illness, panic disorder, or agoraphobia; persistent abnormal illness beliefs (e.g., disease conviction and repeated reassurance-seeking seen in hypochondriacal patients); a pervasive and enduring personality disorder; or a compensation claim resulting from a work-related injury.

Cognitive-behavioural treatment (CBT) is more likely to be effective in those patients with relatively discrete syndromes of short duration (less than five years), for example, chronic fatigue, non-cardiac chest pain, and chronic pelvic pain. This treatment often forms part of a comprehensive, multidisciplinary approach to chronic pain (Philips 1987), and has been shown to be effective in group settings (Skinner *et al.* 1990).

Other patients with relapsing/remitting affective and phobic disorders that present somatically but who function relatively well in between relapses may also be helped by support and judicious use of psychotropic drugs. However, there is a group of patients with long-standing symptoms (e.g., 10–30 years or more) who have pervasive personality disturbances, considerable functional impairment and, in some, an investment in the sick role. To further complicate matters, they often have a coexisting physical disorder. It is not uncommon for these patients to have disturbed backgrounds with histories of parental indifference and/or physical and sexual abuse. They may require more long-term supportive psychotherapy, with encouragement to gently explore psychosocial issues instead of reiterating somatic complaints. The management of such patients is a considerable undertaking, and, in our experience, is best carried out with the co-operation and involvement of the general practitioner (GP). Indeed, there is scope for 'shared care' of these patients, with the psychiatrist and GP providing continuity of support. If a psychiatrist undertakes to manage such patients, then the goals of treatment should be made explicit, i.e., 'coping not curing' as a treatment objective, and out-patient visits should be regularly scheduled (not symptom-contingent).

As this 'intractable' group of chronic somatizers can be very demanding and persistent, limit-setting is essential at the outset. We have therefore developed a number of basic 'ground rules' for the management of chronic somatizers, and these are listed in Table 10. Receptionists, secretaries, and other non-medical staff should be involved in this process (Bass 1990).

Table 10 Ground rules for chronic somatizers

1. Arrange regular spaced visits for specified period, e.g., 6 months.
2. Do *not* see out of hours.
3. No more than one phone call a week.
4. Discuss with secretaries/receptionists.
5. Inform GP, spouse.
6. Keep your own set of notes.
7. Arrange support for yourself.

Treatment studies of patients with chronic somatization

A number of impressive treatment studies have been carried out in the last decade—most of these on relatively homogeneous groups of patients with chronic somatization (see Table 11). None of these studies will be described in great detail. Suffice to say that a wide variety of treatments have been employed, including tricyclic antidepressants (Feinmann *et al.* 1984); a combination of drugs and counselling (Farquhar *et al.* 1989); cognitive-behavioural therapy (Philips 1987; Butler *et al.* 1991; Klimes *et al.* 1990); short-term dynamic psychotherapy (Guthrie *et al.* 1991; Guthrie 1991); and hypnotherapy (Whorwell *et al.* 1984). An innovative treatment for hypo-chondriasis, based on a cognitive-behavioural formulation, is currently being evaluated in a controlled study by Warwick (personal communication).

The results of these studies suggest that a wide variety of treatments are applicable in patients with chronic somatization. All demonstrate the importance of a careful assessment of baseline symptoms and behaviours, and the need for multiple measures of outcome (e.g., subjective pain experience, quantitative measures of pain, behavioural avoidances, functional impairment). One study included a blind assessment of outcome by a physician, which confirmed the efficacy of the intervention (Guthrie *et al.* 1991).

An interesting finding was that dynamic psychotherapy (a treatment not usually associated with successful outcome in somatizing patients) was effective in alleviating symptoms in two-thirds of patients with refractory irritable bowel syndrome (Guthrie 1991). Factors that may have contributed to the success of this treatment included: (1) the initial assessment was carried out in the gastroenterology clinic; (2) a lengthy initial session of psychotherapy lasting 2 to 4 hours (allowing time for the development of a strong therapeutic alliance and aiding the emergence of transference material); (3) the explicit use of bowel symptoms as metaphors to help

Table 11   Characteristics of
chronic somatizers who fail to respond to
psychological treatment

Long history or previous unsuccessful surgery for pain.[a]
Absence of life event before pain onset.[b]
Pain of chronic duration.[a,b]
Constant, unremitting pain.[a]
Pain not aggravated by stress, anxiety.[a]
Absence of anxiety and depression.[a,c]
Treatment-resistant affective disorder.[c]
Strong attribution of symptom(s) to a physical cause.[c]

---

[a] Guthrie *et al.* (1991). [b] Feinmann *et al.* (1984). [c] Butler *et al.*
(1991).

generate a feeling language with the patient (e.g., 'you feel all churned up
inside'). The results of this impressive study should encourage others to
assess the efficacy of psychotherapy in selected groups of patients with
chronic somatization.

Characteristics of patients who fail to respond to psychiatric intervention

This is an important but neglected issue, and it is essential for psychiatrists
and psychologists to be aware of those factors that predict either a poor
response to psychological treatment or refusal to engage in it. Some of these
are summarized in Table 11. Future studies should address these issues in
more detail.

Clinical guidelines

Chronic somatization is common in primary care and general hospital
settings but until recently has attracted insufficient attention from psychiatrists. The following is a summary of treatment strategies.

1. Establishing a diagnosis of chronic somatization is time-consuming and
    requires the collection of a considerable amount of clinical data. Life
    charts are particularly helpful in these patients because they often
    demonstrate a relationship between symptoms and life events and the
    persistence of somatic complaints despite medical intervention (see
    chap. 18). The psychiatrist should discuss each patient's clinical history
    with the GP before embarking on treatment.

2. The most important (and difficult) part of treatment is *engagement*, which requires clinical skills and a supportive, non-adversarial style of interviewing.

3. The content of treatment depends on the type of clinical problem and maintaining factors. For example, patients with chronic pain or chronic fatigue syndrome with stable pre-morbid personalities and adjustment will require different management from those patients with somatization disorder.

4. After establishing a therapeutic alliance, a wide range of therapeutic techniques can be used. These include dynamic psychotherapy, cognitive-behavioural therapy, and drug treatment. Behavioural treatments have been used for many years to treat chronic somatizers in pain clinics.

   Intervention studies have demonstrated the efficacy of psychological treatments, but in future, follow-up periods need to be extended for longer than 12 months.

5. In a sample of chronic somatizers (those satisfying criteria for somatization disorder or Briquet's syndrome) 'damage limitation' is a more realistic therapeutic goal than cure. The psychiatrist should explain empathically that cure is unlikely, but convey hope that the treatment can help the patient live a better life.

6. The management of chronic somatizers offers the psychiatrist an opportunity to collaborate with both primary care doctors and hospital physicians. As the prevalence of somatization is so high (4.4 per cent in some community studies) there is also considerable scope for collaborative research in this field.

References

APA (American Psychiatric Association) (1980). *Diagnostic and statistical manual of mental disorders* (3rd edn). Washington DC.

APA (American Psychiatric Association) (1987). *Diagnostic and statistical manual of mental disorders* (3rd edn, revised). Washington DC.

Bass, C. M. (1990). Assessment and Management of patients with Functional Somatic Symptoms. In *Somatization: physical symptoms and psychological illness* (ed. C. Bass), pp. 40–72. Blackwell, Oxford.

Bass, C. M. and Murphy, M. R. (in prep.). A case control study of British patients with somatization disorder (SD)

Benjamin, S. *et al.* (1988). The relationship of chronic pain, mental illness and organic disease. *Pain*, **32**, 185–95.

Bhatt, A., Thomenson, B., and Benjamin, S. (1989). Transcultural patterns of somatization in primary care: a preliminary report. *Journal of Psychosomatic Research*, **33**, 671–80.

Bird, J. and Cohen-Cole, S. A. (1990). The Three-Function Model of the Medical Interview. An educational device. In *Advances in psychosomatic medicine*, Vol. 20. *Methods in teaching consultation/liaison psychiatry* (ed. M. S. Hale), pp. 65–88. Karger, Basel.

Bridges, K. W. and Goldberg, D. P. (1985). Somatic presentation of DSM III psychiatric disorders in primary care. *Journal of Psychosomatic Research*, **29**, 563–9.

Butler, S., Chalder, T., Ron, M., and Wessely, S. (1991). Cognitive behaviour therapy in chronic fatigue syndromes. *Journal of Neurology, Neurosurgery & Psychiatry*, **54**, 153–8.

De Leon, J., *et al.* (1987). Why do some psychiatric patients somatize? *Acta Psychiatrica Scandinavica*, **76**, 203–9.

Escobar, J. I., *et al.* (1987). Somatization in the community. *Archives of General Psychiatry*, **44**, 713–18.

Farquhar, C. M., Rogers, V., Franks, S., Pearce, S., Wadsworth, J., and Beard, R. W. (1989). A randomised controlled trial of medroxyprogesterone acetate and psychotherapy for the treatment of pelvic congestion. *British Journal of Obstetrics & Gynaecology*, **96**, 1153–62.

Feinmann, C., Harris, M., and Cawley, R. (1984). Psychogenic facial pain: presentation and treatment. *British Medical Journal*, **288**, 436–8.

Gask, L., Goldberg, D., Porter, R., and Creed, F. (1989). Treatment of somatisation: evaluation of a teaching package with general practice trainees. *Journal of Psychosomatic Research*, **33**, 697–703.

Goldberg, D., Gask, L., and O'Dowd, T. (1989). The treatment of somatization: teaching techniques of reattribution. *Journal of Psychosomatic Research*, **33**, 689–95.

Guthrie, E. (1991). Brief psychotherapy with patients with refractory irritable bowel syndrome. *British Journal of Psychotherapy*, **8**, 175–88.

Guthrie, E., Creed, F., Dawson, D., and Tomenson, B. (1991). A controlled trial of psychological treatment for the irritable bowel syndrome. *Gastroenterology*, **100**, 450–7.

House, A. (1990). Hypochondriasis and related disorders. Assessment and management of patients referred for a psychiatric opinion. *General Hospital Psychiatry*, **11**, 156–65.

Katon, W. *et al.* (1992). A randomised trial of psychiatric consultation with distressed high-utilisers. *General Hospital Psychiatry*, **14**, 86–98.

Kirk, J. (1989). Cognitive-behavioural assessment. In *Cognitive behaviour therapy for psychiatric problems: a practical guide* (ed. K. Hawton, P. M. Salkovskis, J. Kirk, and D. M. Clark), pp. 13–51. Oxford University Press.

Kleinman, A. and Kleinman, J. (1985). Somatization: the interconnections in Chinese society among culture, depressive experiences and the meaning of pain. In *Culture and depression* (ed. A. Kleinman and B. Good), pp. 429–90. University of California Press, Berkeley.

Klimes, I., Mayou, R. A., Pearce, M. J., Coles, L., and Fagg, J. (1990). Psychological treatment for atypical non-cardiac chest pain: a controlled evaluation. *Psychological Medicine*, **20**, 605–11.

Mezzich, J. E. (1988). On developing a psychiatric multiaxial scheme for ICD-10. *British Journal of Psychiatry*, **152**(Suppl. 1), 38–44.

Parker, G., Tupling, H., and Brown, L. B. (1979). A parental bonding instrument. *British Journal of Medical Psychology*, **52**, 1–10.

Philips, C. (1987). The effects of behavioural treatment on chronic pain. *Behaviour Research and Therapy*, **25**, 365–78.

Pilowsky, I. (1967). Dimensions of hypochondriasis. *British Journal of Psychiatry*, **113**, 89–93.

Robins, L. N., Helzer, J. E., Croughan, J., and Ratcliff, J. S. (1981). National Institute of Mental Health, Diagnostic Interview Schedule: its history, characteristics, and validity. *Archives of General Psychiatry*, **38**, 381–9.

Rosen, G., Kleinman, A., and Katon, W. (1982). Somatization in family practice: a biopsychosocial approach. *Journal of Family Practice*, **14**, 493–502.

Salkovskis, P. M. (1989). Somatic Problems. In *Cognitive behaviour therapy for psychiatric problems: a practical guide* (ed. K. Hawton, P. M. Salkovskis, J. Kirk, and D. M. Clark), pp. 235–76. Oxford University Press.

Skinner, J. B., Erskine, A., Pearce, S., Rubinstein, I., Taylor, M., and Foster, C. (1990). The evaluation of a cognitive behavioural treatment programme in outpatients with chronic pain. *Journal of Psychosomatic Research*, **34**, 13–20.

Smith, R. G., Monson, R. A., and Ray, D. C. (1986). Psychiatric consultation in somatization disorder. *New England Journal of Medicine*, **314**, 1407–13.

Swartz, M. *et al.* (1986). Somatization disorder in a community population. *American Journal of Psychiatry*, **143**, 1403–8.

Whorwell, P. J., Prior, A., and Faragher, E. B. (1984). Controlled trial of hypnotherapy in the treatment of severe refractory irritable bowel syndrome. *Lancet*, **ii**, 1232–4.

10

Factitious disorders: what can the psychiatrist do?

MICHIEL W. HENGEVELD

Introduction

Patients with a factitious disorder (FD) present their physicians and psychiatrists with many difficulties and dilemmas. Their illness is often not recognized and very difficult to establish. The physician's response to such patients is often at first fascination by the intriguing symptoms, then bewilderment at the incongruous findings, and finally contempt and anger at having been tricked. The thought of patients feigning an illness is difficult to understand and hard to accept. The treatment of FD patients is often interfered with by the strong and negative emotions they evoke in their physicians.

Diagnosing FD poses the physician with the dilemma of believing the patient, for the sake of the patient–doctor relationship, or catching the patient in a lie or in the act of producing symptoms, in order to establish diagnostic certainty. Another dilemma is whether or not patients should be confronted with the true nature of their illness. A third dilemma is that the patient with factitious physical symptoms wishes to be treated by the physician but that the physician would rather refer these patients to the psychiatrist. The psychiatrist, on the other hand, cannot treat the patient without the involvement of the physician. The management of FD, there- fore, requires a strategy that is planned collaboratively between physician and psychiatrist.

General characteristics

Factitious disorders are characterized by symptoms or signs that are inten- tionally feigned or produced. The symptoms and signs can be manifested at three levels: (1) fictitious history (as in supplying the classic history for intermittent porphyria); (2) feigned presentation (e.g., complaints of typical renal colic pain); (3) or self-inflicted pathophysiology (as in the production of painful swellings in the muscles by injection of paraffin).

The judgement that the symptoms or signs are intentionally produced or feigned can only be inferred by the physician, based on the likelihood that the patient has fabricated the history or complaints, absence of objective signs to explain the symptoms, an unusual clinical picture that can be understood from the patient's concept of the illness, or detection of the means used by the patient to inflict the signs of the disorder. The production of symptoms is not voluntary in the strict sense of the word, because it has a certain compulsive character, and the patient is unable to refrain from it. Probably the self-mutilating behaviour generally takes place with full consciousness, although in some cases it may happen in a kind of dissociative state.

The behaviour of the FD patient does not seem to be motivated by any external gains, but can only be explained by assuming a psychological need to adopt the sick role.

Table 12   Types of factitious disorders

Acute abdominal	Laparotomophilia migrans
Cardiac	Feigned myocardial infarction
Cutaneous	Dermatitis artefacta
Endocrinological	Factitious diabetes mellitus
Gastrointestinal	Laxative abuse, self-induced vomiting
Haemorrhagic	Haemorrhagia histrionica
Infectious	Hyperpyrexia figmentatica
Neurological	Neurologica diabolica, factitious epilepsy
Pulmonary	Factitious haemoptysis
Urogenital	Factitious haematuria

Subtypes

Most medical and surgical specialities harbour their own type of factitious patients, often baptized with colourful, pseudo-scientific names (Asher 1951; Spiro 1968; Ford 1983*a*). Some of these are shown in Table 12. As Spiro (1968) formulated: 'This nomenclature is perhaps symptomatic of the mixture of bemusement, bewilderment, contempt, and anger that these patients arouse in their physicians'. A more simple classification is presented in DSM-III-R (APA 1987), which distinguishes three main types: (1) factitious disorder with psychological symptoms; (2) factitious disorder with physical symptoms; and (3) factitious disorder not otherwise specified.

Factitious disorder with psychological symptoms   This is more difficult to diagnose than FD with physical complaints, as there is no way of excluding

a 'true' psychiatric disorder by physical examination or laboratory investigations. The disorder may be recognized by the pan-symptomatic complex
or typical textbook qualities of psychological symptoms that are presented
and by the fact that the symptoms are worse when the patient is aware of
being observed. The patient may claim depression and suicidal ideation
following the alleged and often atrocious death of a partner, memory loss,
hallucinations, or dissociative and conversion symptoms (American Psychiatric Association 1987). Sussman and Hyler (1985) published a short, but
comprehensive textbook review. Factitious psychotic symptoms may be
more common than generally realized, and indicate a poor prognosis (Pope
et al. 1982). Patients presenting with feigned bereavement or depression
more often show the characteristics of a Munchausen syndrome, and may
therefore be managed the same way. This chapter will mainly confine itself
to the management of FD with physical symptoms.

Table 13   Typical features of Munchausen's syndrome

Feigned severe illness of an urgent nature.
Dramatic, but plausible history (often part truth).
Pathological lying (pseudologia fantastica).
Spurious signs of disease produced by self-mutilation or drug abuse.
Willingness to undergo physical investigation and treatment.
Evidence of many previous hospital procedures (e.g., surgical scars).
History of admissions to many hospitals.
Peregrination (travelling around).
Aggressive, disruptive or evasive behaviour, once admitted to hospital.
Absence of visitors during hospital stay.
Premature self-discharge from hospital against medical advice.
Absence of any readily discernible ulterior motive.

Factitious disorder with physical symptoms   This is often subdivided into
two subtypes, namely those with a short history and those which are
chronic. Although chronic FD may not be the most prevalent subtype of the
FDs, it has gained greatest attention in the literature, probably because of
its dramatic and anger-provoking nature (Reich and Gottfried 1983). It was
Asher (1951) who named this illness after the famous Baron von Munch-
hausen, who travelled widely and allegedly told theatrical and untruthful
stories. The features that may be found in typical cases (Stone 1977; Aduan
et al. 1979; Hyler and Sussman 1981) are presented in Table 13.

Reich and Gottfried (1983), however, surveyed 41 FD in-patients identified in their teaching hospital during a 10 year period and found that most

cases did not show these characteristics. Although several Munchausen patients visited the emergency wards during the survey period, none of them were admitted to the hospital. They divided the 41 FD patients into four subgroups according to the method used to produce the illness: self-induced infections; simulated illnesses; chronic wounds; and surreptitious self-medication. These subgroups differed in terms of the chronicity and severity of the disorders, and the personality characteristics and attitudes of the patients. They also had implications for management and prognosis. Eisendrath (1984) suggests that the 'level of enactment' (fictitious history, feigned presentation of signs, or creation of verifiable abnormalities) and the extent of recurrence determine treatment choice and outcome. Another distinction has been made by Kooiman (1987) between FD patients with or without an early disturbance in personality development, implying that this has implications for the management of the patient.

Differential diagnosis

Factitious disorder is part of a spectrum of psychiatrically explained 'physical non-diseases'; at one end of the spectrum are the *somatoform disorders* like hypochondriasis, conversion disorder and somatization disorder (see Chapter 9); at the other end is *malingering*. There is a considerable similarity and overlap between these disorders; moreover, they can occur simultaneously in one patient. All are characterized by the 'choice of the medical field as the stage for playing out their conflicts' (Cramer *et al.* 1971), by the need to be sick and to submit to hospital interventions and by the absence of objective findings to explain the symptoms (Glickman 1980). In somatoform disorders, however, the symptoms and signs are not produced intentionally, there is generally less sophistication in the knowledge of medical terminology and there may be a direct temporal relation or a symbolic reference to a specific emotional conflict or an environmental stressor (Hyler and Sussman 1981; Sussman and Hyler 1985). In malingering, there is an obvious, recognizable environmental goal in producing the symptoms and the patients are aware of the methods they use and of the gain (Glickman 1980; Sussman and Hyler 1985).

Epidemiology

Very little is known about the prevalence, age at onset, sex ratio, and course of FD. Some believe it is very rare, others that it is more common than generally believed because it often remains undetected. Munchausen patients may be over-reported, because of their clinical interest and because of the existence of multiple accounts of one patient (Hyler and Sussman 1981). Psychiatric consultation-liaison services report that FD is diagnosed in 0.2–1.5 per cent of referrals (Kooiman 1987; Hengeveld *et al.* 1988; Sutherland and Rodin 1990). The age of onset is generally before middle age, cases being described with an age range from 17 to 72 years and an average age of 30 years (Kooiman 1987). In an unpublished report on 8509 psychiatric consultations in eight hospitals in The Netherlands, the mean age of the 45 FD patients was 34.3 years, the percentage of women was 69 per cent, and most patients were single and unemployed. The typical Munchausen patients are probably more often male (Ford 1983*b*). The prognosis of the illness is generally described as gloomy or poor; however, no systematic studies have been carried out to test this clinical impression.

Aetiology

Although, for obvious reasons, studies of larger groups of patients are lacking, some typical features seem to characterize the history of many FD patients (Spiro 1968; Hyler and Sussman 1981), namely:

- parental abuse, neglect or abandonment;
- early experiences with chronic illness or hospitalization;
- a significant relationship with a physician in the past;
- experiences of medical mismanagement; and
- employment in the medical field.

Personality traits typical for DSM-III-R borderline personality disorder (APA 1987) are often mentioned in the literature (Hyler and Sussman 1981; Ford 1983*b*). This may be particularly true only for the typical Munchausen cases. In an unpublished Dutch consultation-liaison report a DSM-III-R personality disorder was diagnosed in 79 per cent of the 45 FD patients.

Based on these characteristics, psychodynamic theories of FD generally contain the following themes (Spiro 1968; Cramer *et al.* 1971; Ford 1983*b*; Eisendrath 1984; Sussman and Hyler 1985):

- deprivation and neglect, preventing patients from achieving a necessary sense of acceptance;
- dependency, the hospital and its staff becoming a surrogate home and family, illness bringing reward instead of suffering;
- partial suicide, the self-mutilation being equated with introjected aggression and with an act of revenge;
- masochism, interpreted as an attempt to relieve unconscious guilt (e.g., about the anger about being deprived);
- occupational association, suggesting an oscillation between identification with the physician and with the patient role; and
- mastery of the childhood illness experiences by reliving them and by controlling them through intentional simulation.

Most psychoanalytically oriented authors envisage the pathomimicry as a last-ditch attempt to ward off further psychological disintegration (Cramer *et al.* 1971; Ford 1983*b*).

Management

No systematic studies have been performed into the results of treatments of FDs, except for a recent comparison by Van Moffaert (1991) of two management approaches of a large, heterogeneous group of patients with self-inflicted cutaneous lesions, the majority of whom were classified as having dermatitis artefacta. Therefore, the therapy recommendations presented below are based on case reports and clinical experience.

Confrontation

Earlier authors recommended vigorous confrontation of patients with the factitious nature of the illness (Stone 1977; Aduan *et al.* 1979; Ford 1983*a,b*). However, caution was advised in the case of psychotic or very socially unskilled patients (Stone 1977). Some authors suggested that the psychiatrist should confront the patient preferably after transfer to a locked psychiatric ward (Stone 1977; Ford 1983*b*); others preferred the physician or surgeon to do this, in a calm, matter-of-fact, non-judgemental manner, prior to psychiatric consultation (Ford 1983*a*, Reich and Gottfried 1983; Eisendrath 1984). The patients' reaction to confrontation, however, often includes denial, anger, and leaving hospital against advice (Ford 1983*a*). Nowadays, most authors warn against early confrontation of the patient, as it often fails and sets up the patient as an adversary (Sussman and Hyler 1985; Eisendrath 1989; Van Moffaert 1991). In her study on 159 dermatitis artefacta patients, Van Moffaert (1990) found a better prognosis when confrontation was avoided in the early treatment phase, and that complete remission of the self-inflicted physical lesions and of the psychiatric disorder could be achieved without any confrontation. Only after establishment of a good relationship with the patient, careful confrontation might be considered, e.g., in patients without severe personality disorder (Reich and Gottfried 1983; Eisendrath 1989; Van Moffaert 1991).

In-patient psychiatric treatment

Yassa (1978) was the first to describe successful treatment of a case of Munchausen's syndrome. The patient had been hospitalized almost continuously over a period of 12 years, escaping many times to present herself with FD to general hospitals. It was then decided to treat her with a behaviour modification programme, consisting of rewarding positive attributes, negatively reinforcing unwanted behaviour by denying privileges, and giving her supportive therapy in 30 minute sessions once a week. The programme lasted three years, leading to discharge from the hospital, working in a steady job, and discontinuation of feigned illness.

The in-patient therapy of two cases of chronic FD with both physical and psychiatric symptoms was described by O'Shea and McGennis (1982). After observation and analysis of the patients' behaviour, a hierarchy of re-inforcers of Munchausen behaviour was constructed. It appeared that touch, emotional display, and use of humour were far more effective than verbal psychotherapy. The treatment involved non-judgemental acceptance of the patient's history and fostering of a sibling-type relationship with the therapist. An initial deep dependency was tolerated and graduated individuation was encouraged and praised. As a result of this strategy the FD behaviour of both patients almost completely ceased. However, they remained in the psychiatric hospital.

Outpatient treatment

Klonoff *et al.* (1983–4) described the behavioural approach to a young woman with chronic FD and so-called borderline personality features. Treatment goals were specified as: (1) to increase her sense of control and self-esteem; (2) to help her establish better interpersonal relationships; (3) to promote age-appropriate social and occupational behaviours; and (4) to diminish her self-destructive behaviour. Her tendency to split people into 'good' or 'bad' was capitalized on by letting two therapists play different roles: the 'good therapist', responsible for her day-to-day treatment, and the 'bad doctor', responsible for psychiatric issues and interventions with the parents. Over time, these roles were gradually merged. All interventions were structured as much as possible to allow the patient to feel she was in control. Biofeedback was used to change her preoccupation with physical functioning in a positive direction and to enable her to give up symptoms without loosing face. Traditional operant procedures were applied to stop the reinforcement of factitious symptoms. Specifically, both the patient and her family were directed to ignore episodes of 'seizure' activity. Any mention made by the patient of either new or different physical symptoms was systematically ignored by the therapist. Paradox was induced by expressing initial scepticism towards any progress the patient reported, but decreases in symptoms and attempts to increase self-esteem and control of her life were enthusiastically reinforced, primarily by the 'good' therapist. The patient's physical symptoms almost disappeared, the number of emergency room visits decreased substantially, her social functioning improved, and she successfully resumed her college studies.

A non-confrontational approach to chronic FD patients (excluding typical Munchausen patients with antisocial traits) is described by Eisendrath (1989). To alter the patient's behaviour he uses a 'double-bind' in the following manner: the clinician presents the patient with a differential diagnosis of the disorder, indicating the possibility of FD. Following such contradictory interpretations of the illness, the patient has the option of responding to medical care or confirming the presence of a FD. (For example, the differential diagnosis presented to a patient with wilful interference with wound-healing included the possibility of an unusual wound-healing problem that should respond to one more graft attempt. If it did not, a diagnosis of factitious disorder would be established. The graft took place, and there was no recurrence of infection at two-year follow-up.)

Eisendrath (1989) also utilizes so-called 'inexact interpretations' to suggest a relationship between certain events or emotional stressors (e.g., feeling abandoned) and the emergence of symptoms. This technique involves a statement that captures much of the psychodynamics of the patient's problem without specifically identifying the factitious cause of the disorder. (For example, a male patient with recurrent scrotal inflammation appeared to have applied phenol solution to his scotum after he had had sexual encounters with prostitutes. Usually these liaisons occurred after he had felt rejected by a woman in his church. Without describing the factitious nature of the symptoms, the psychiatrist suggested that the patient's feelings about his perceived rejections and forbidden liaisons could be affecting his scrotum.)

Finally, face-saving techniques such as biofeedback and hypnosis are provided, so that patients can relinquish their symptoms without exposure and humiliation.

Liaison approach

In the recent literature, more emphasis is placed on the integration of the medical/surgical and psychiatric management of FD patients with physical symptoms. As these patients, in general, present themselves to physicians and surgeons, and are often not motivated to enter psychiatric treatment, this seems to be a more realistic approach.

Factitious disorder patients, especially those with Munchausen's disorder, mobilize the physician's interest by displaying suffering and helplessness and by presenting the picture of an unusual and challenging diagnostic problem. They may even succeed in seducing the surgeon to operate on them. However, when the hoax is uncovered, the medical or surgical staff respond with angry condemnation of the patient (Cramer *et al.* 1971). The chief stumbling block in the initial management of the FD patients lies, therefore, in the negative feelings elicited in general hospital staff (Stone 1977). Hyler and Sussman (1981) and Van Moffaert (1991) extensively describe how the liaison psychiatrist may prevent the development of such destructive and stressful situations for both the patient and the hospital staff. They suggest the psychiatrist should: (1) explain the nature and dynamics of the disorder to the staff; (2) discuss their suspicions, frustration, and hostility; (3) address the patient's splitting of the medical and nursing staff; and (4) avoid confronting the patient with the factitious nature of the symptoms.

Van Moffaert (1991) compared general hospital in- or out-patient liaison treatment of 208 (159 FD) patients with self-inflicted cutaneous lesions, with psychiatric in- or out-patient treatment of 37 (9 FD) patients with the same type of disorder (but with more severe psychiatric co-morbidity). She found a much higher rate of patient compliance (76 per cent vs. 44 per cent) and complete healing of lesions (50 per cent vs. 38 per cent) in the liaison group. She, therefore, recommends that FD patients be managed by general hospital physicians and surgeons, supported by a liaison psychiatrist. Key components of this integrative treatment are:

- psychotropic drugs for (atypical) depression, anxiety, impulsivity, and other personality disorder symptoms (see Cowen 1990);
- palliative medical and surgical measures (e.g., surgical dressing of a wound, occlusive bandaging, ointments or a placebo drug) as a form of supportive therapy (symbolizing the medical concern and care these patients are basically craving for);
- avoidance of fostering extreme dependency needs, leading to medical or surgical over-consumption;
- avoidance of a mistrustful attitude during nursing procedures;
- avoidance of a triumphant reaction when the self-inflicted nature of the lesions is proven;
- discussion of the events that appear to trigger symptom production;
- ventilation of the patient's anger in order to help dispel aggressive feelings; and
- an accepting and tolerant attitude on the part of the physician, thus providing a corrective emotional experience for the patient.

Choice and goal of treatment

The therapies described above are, of course, not all comparable as far as indication, type of patient, setting, and investment of time and energy are concerned. A chronic FD patient with severe antisocial personality traits will be treated differently from a patient with a stable family and work life presenting with one episode of FD. In general, the liaison approach will be the first and most obvious strategy chosen. In the typical Munchausen patient, the only aim of this approach may be to forestall the patient from aggressive diagnostic and therapeutic procedures and to reduce the emotional and financial burden for general hospital staff (Hyler and Sussman 1981; Ford 1983*b*). In the less severe cases, the accomplishment of a continuing, non-critical doctor–patient relationship may lead to the cessation of FD behaviour and to psychosocial readjustment. If the patient can be motivated for psychiatric out-patient therapy, non-confrontational techniques can be applied to diminish the feigning of symptoms. Other problem areas found in FD patients (e.g., substance abuse, eating disorders, and sexual problems) can be approached with a variety of cognitive-behavioural techniques (Hawton *et al.* 1989). FD patients with very disabling personality disorders, often feigning both physical and psychological symptoms, may have to be treated in a psychiatric hospital, applying the more classical behavioural techniques described above. In the unpublished Dutch consultation-liaison report, 44 per cent of the 45 FD patients were referred for psychiatric aftercare, and only 4 per cent were transferred to a psychiatric hospital.

Clinical guidelines

1. FD can generally only be inferred by the physician, based on a conflicting history or an unusual clinical picture. The symptoms and signs can only be understood by assuming a psychological need to adopt the patient role. A history of early experiences with severe illness or hospitalization and close professional or personal connections with physicians may support the diagnosis of FD.

2. FD patients often do not exhibit the typical features of Munchausen's syndrome and borderline personality disorder. FD may occur in a single episode, may be chronic but without severely pathological personality traits, and may be mixed with psychological factitious symptoms.

3. FD may also occur simultaneously with somatoform disorders or with malingering; there is a considerable overlap between these disorders.

4. Confrontation of the patient with the factitious nature of the illness should often be avoided, certainly in the early treatment phase. Assuming a psychogenesis of the symptoms and signs, without explicitly suggesting the factitious nature of the disorder, may be a way of coming to some agreement with the patient and also with the patient's relatives.

5. A psychiatric liaison approach is probably the most effective (initial) treatment of all FD patients with physical symptoms. The psychiatrist should explain the patient's illness and help the physician and nurses deal with their counter-productive negative emotions in order to avoid angry confrontation and disruption of the doctor–patient relationship. Medical, surgical, and nursing care, without fostering the patient's dependency too much, may temporarily support the patient's need for being ill. Some form of out-patient medical or surgical care should be continued, including when the patient has been referred to the psychiatrist.

6. In psychiatric treatment, suggestive techniques, such as paradoxical intention ('This illness will probably not get better') or double-bind interpretations ('If the illness does not get better, it must be FD'), operant techniques, such as ignoring the symptoms and reinforcing

healthy behaviour, or psychodynamic techniques, such as discussing the supposed psychological mechanisms leading to the onset or perpetuation of the FD can be applied to diminish the feigning of symptoms. The relatives of the patient may sometimes be instructed to ignore the factitious behaviour.

7. Only after the establishment of a good therapist–patient relationship should careful confrontation of the patient with the factitious nature of the illness be considered, e.g., in patients without severe personality disorders.

References

Aduan, R. P., Fauci, A. S., Dale, D. C., Herzeberg, J. H., and Wolff, S. M. (1979). Factitious fever and self-induced infection: a report of 32 cases and review of the literature. *Annals of Internal Medicine*, **90**, 230–42.

APA (American Psychiatric Association) (1987). *Diagnostic and Statistical Manual of Mental Disorders (3rd edn, revised)*. American Psychiatric Association, Washington, DC.

Asher, R. (1951). Munchausen's syndrome. *Lancet*, **i**, 339–41.

Cowen, P. J. (1990). Personality disorders: are drugs useful? In *Dilemmas and difficulties in the management of psychiatric patients* (ed. K. Hawton and P. J. Cowen), pp. 105–16. Oxford University Press.

Cramer, B., Gershberg, M. R., and Stein, M. (1971). Munchausen syndrome: its relationship to malingering, hysteria, and the physician–patient relationship. *Archives of General Psychiatry*, **24**, 573–8.

Eisendrath, S. J. (1984). Factitious illness: a clarification. *Psychosomatics*, **25**, 110–17.

Eisendrath, S. J. (1989). Factitious physical disorders: treatment without confrontation. *Psychosomatics*, **30**, 383–7.

Ford, C. V. (1983a). Factitious illness. In *The somatizing disorders*, pp. 135–54. Elsevier, New York.

Ford, C. V. (1983b). The Munchausen syndrome. In *The somatizing disorders*, pp. 154–75. Elsevier, New York.

Glickman, L. S. (1980). *Psychiatric consultation in the general hospital*, pp. 170–4. Dekker, New York.

Hawton, K. E., Salkovskis, P. M., Kirk, J., and Clark, D. M. (ed.) (1989). *Cognitive behaviour therapy for psychiatric problems: A practical guide*. Oxford University Press.

Hengeveld, M. W., Huyse, F. J., Mast, R. C. van der, and Tuinstra, C. L. (1988). A proposal for standardization of psychiatric consultation-liaison data. *General Hospital Psychiatry*, **10**, 410–22.

Hyler, S. E. and Sussman, N. (1981). Chronic factitious disorder with physical symptoms (the Munchausen syndrome). In *The medically ill patient* (ed. J. J. Strain), pp. 365–77. Saunders, Philadelphia.

Klonoff, E. A., Youngner, S. J., Moore, D. J., and Hershey, L. A. (1983–4). Chronic factitious illness: a behavioral approach. *International Journal of Psychiatry in Medicine*, **13**, 173–83.

Kooiman, C. G. (1987). Neglected phenomena in factitious illness: a case study and review of the literature. *Comprehensive Psychiatry*, **28**, 499–507.

O'Shea, B. and McGennis, A. (1982). The psychotherapy of Munchausen's syndrome. *Irish Journal of Psychotherapy*, **1**, 17–19.

Pope, H. G., Jonas, J. M., and Jones, B. (1982). Factitious psychosis: phenomenology, family history, and long-term outcome of nine patients. *American Journal of Psychiatry*, **139**, 1480–3.

Reich, P. and Gottfried, L. A. (1983). Factitious disorders in a teaching hospital. *Annals of Internal Medicine*, **99**, 240–7.

Spiro, H. R. (1968). Chronic factitious illness. *Archives of General Psychiatry*, **18**, 569–79.

Stone, M. H. (1977). Factitious illness: psychological findings and treatment recommendations. *Bulletin of the Menninger Clinic*, **41**, 239–54.

Sussman, N. and Hyler, S. E. (1985). Factitious disorders. In *Comprehensive textbook of psychiatry* (ed. H. I. Kaplan and B. J. Sadock), pp. 1242–7. Williams & Wilkins, Baltimore.

Sutherland, A. J. and Rodin, G. M. (1990). Factitious disorders in a general hospital setting: clinical features and a review of the literature. *Psychosomatics*, **31**, 392–9.

Van Moffaert, M. (1990). Self-mutilation: diagnosis and practical treatment. *International Journal of Psychiatry in Medicine*, **20**, 373–82.

Van Moffaert, M. (1991). Integration of medical and psychiatric management in self-mutilation. *General Hospital Psychiatry*, **13**, 59–67.

Yassa, R. (1987). Munchausen's syndrome: a successfully treated case. *Psychosomatics*, **19**, 242–3.

11

Personality disorders: do psychological treatments help?

PATRICIA R. CASEY

Introduction

The pessimism concerning the treatment of personality disorders is pervasive and is reflected in current clinical practice by the removal of this group of disorders from the Mental Health Act, in so far as treatability is concerned. Even amongst those who profess a modicum of optimism there is little empirical information about treatment choice and response.

Several factors contribute to this therapeutic nihilism. Foremost is the inadequacy and muddled thinking about the diagnosis and classification of personality disorders. There has been a notable inability among psychiatrists to separate state symptoms from long-standing traits despite the fact that this is an obvious clinical exercise. Thus, patients with depressive illness who exhibit apathy and anxiety are frequently incorrectly adjudged to have an underlying personality disorder. Moreover, the negative cognitions of those with depressive illness often leads them to describe negative traits in themselves—a problem which can be overcome by simply interviewing a reliable informant. Elementary as these rules may be they are frequently flouted and contribute to the poor reliability of the diagnosis. In addition, until recently there were no adequate definitions for the various personality disorders and expressions, such as 'inadequate', 'immature', and a host of others have abounded with no agreement on what these mean. In many instances these labels have been no more than a pejorative and unhelpful description of a difficult patient. The counter approach, namely to operationalize every cluster of abnormal traits, has perhaps led to its own constellation of problems, not least of which is the excessive proliferation of categories of seemingly scientific merit but which have not been validated and may be no more than the 'flavour of the month'. Since, in most psychiatric disorders, treatments were developed after the conditions had been adequately defined and were readily diagnosable it is hardly surprising that the treatment of personality disorder has seldom been studied in common clinical settings. The addition of the multi-axial classification in ICD 10 (as in DSM-III-R) (see Table 14) should henceforth assist the profession in conceptualizing more clearly the nature of the condition to be treated, although the burgeoning of categories of uncertain validity is a cause for grave concern and may vitiate any progress in the investigation of therapies for personality disorder.

The second reason for this therapeutic gloom lay in the tradition of equating personality disorder, almost exclusively, with psychopathy—the personality disorder which is generally regarded, even by the most optimistic of therapists, as having a poor prognosis.

The 'collective unconscious' of the psychiatric profession also raises its head in this area. Doctors are trained to effect either 'cure' or maximum improvement in their patient's condition. The prospect of perhaps bringing about relief in only some aspects of the patient's life (e.g., social functioning, coping skills, etc.) is anathema to our 'upbringing' and shunned—although on such limited goals should the success of the psychological treatments for this group of disorders be based.

The critics of the concept of personality disorder have been vociferous in their condemnation on the grounds of the value judgements inherent in it. This is a specious concern which has been articulated and effectively answered in relation to the whole of mental illness. However, a more valid criticism is the use of categories of personality disorder which have not been shown to have heuristic, predictive or clinical value. Despite the limitations of this approach to the conceptualization of personality and despite the pleas for an alternative method, in particular a dimensional one, the traditional nosology persists. So also do approaches to treatment which still revolve round these unsound categories, placing another impediment in the way of a coherent assessment of intervention.

Finally, many of the therapies that are offered to those with personality disorders are extremely time-consuming, require additional lengthy training, and tend to be of uncertain effectiveness despite the grandiose claims of their adherents. In particular, psychodynamic psychotherapy and more recently cognitive therapy can be described as 'catch all' therapies which in this regard are often based on anecdote and case reports. Such approaches are discouraging in the harsh scientific world of psychiatry as we approach the end of the second millenium.

Why should personality disorders be treated?

It is pertinent to pose this question in view of the absence of convincing evidence for the treatability of disorders of personality. However, the answer to this lies in the now convincing body of opinion that not only does personality disorder lead to interpersonal problems and human suffering but also has an impact on the outcome of all psychiatric disorders, whether belonging to those conventionally described as neurotic or psychotic.

Among patients with personality disorder service utilization is also increased (Tyrer and Sievewright 1988) and response to physical treatments, such as antidepressants (Tyrer *et al*. 1983) and ECT (Zimmerman *et al*. 1986; Casey and Butler in prep.), is compromised. The treatment of personality disorders is of more than ephemeral interest to an eccentric group of psychiatrists and psychotherapists and should be the concern of the health service consultant as much as the private therapist.

Natural history of personality disorders

Before considering psychological treatments it is important to examine the course of personality disorders in order to place various treatments in their context and to allow for natural 'recovery' in measuring the success of these. Epidemiological studies provide some clues. In general, personality

Table 14 Suggested division between mature and immature personality disorders according to three diagnostic schemes, showing equivalent disorders each scheme

International Classification of Disease (ICD 10)	Diagnostic and Statistical Manual (DSM-III-R)	Personality Assessment Schedule (PAS)
Mature personality disorders		
Anankastic	Obsessive compulsive	Anankastic
–	–	Hypochondrial
–	–	Anxious
–	–	Asthenic
Paranoid	Paranoid	Paranoid
Schizoid	Schizoid	Schizoid
–	Schizotypal	–
–	–	Dysthymic
Immature personality disorders		
Dyssocial	–	Sociopathic
Dyssocial	Antisocial	Sensitive-aggressive
Impulsive	Borderline	Explosive
Histrionic	Histrionic	Histrionic
Dependent	Dependent	Passive-dependent
Anxious	Avoidant	Avoidant
–	Passive-aggressive	–
–	Narcissistic	–

Adapted from Tyrer (1988*a*) with permission of the publishers.

disorders are more common in the young than in the elderly, suggesting that maturation does occur. In particular, antisocial and borderline personality are rarely diagnosed in those over the ages of 45 and 55 respectively and rates for the antisocial type are highest in the 24 to 44 years age group. There is no information on the age-specific rates for other categories of abnormal personality despite evidence that they can be diagnosed in community and general practice populations in substantial numbers.

There is thus some evidence, from epidemiology as outlined above, but also from statistical analyses, that personality disorders can be divided into those which persist into old age with little variation and those which predominate in the young but attenuate with advancing years. These have been described respectively as the 'mature' and 'immature' personality disorders (Tyrer 1988*a*) and are shown in Table 14. The inclusion of borderline personality in the category that improves with increasing age is in direct contrast to the view of Millon who states: 'Nothing but the most intensive therapy will produce substantial change' (Millon 1981).

Treatments aimed at specific traits and constellations

As there is no conclusive evidence that there are category-specific therapies this section will outline approaches that home-in on specific areas of difficulty, irrespective of the assigned category of abnormal personality. In this way it is hoped to avoid some of the criticisms of therapeutic meaninglessness which have been aimed at the present categorical approach. In general, treatments tend to be behavioural in approach although they have not been tested with any methodological rigour.

Assertiveness and confidence

Social skills training was initially developed as part of rehabilitation programmes for those suffering from schizophrenia but its use has become widespread and reached other groups of patients, in particular those who are socially anxious, lacking confidence and withdrawn. As this constellation of traits may also be associated with difficulties in assertion, assertiveness training is often combined (Trower *et al.* 1978). This deals with the issue of saying 'no', the appropriate expression of anger and the setting of limits to other people's behaviour. Exposure and desensitization may also be applied as an adjunct to these techniques in those who are socially avoidant and anxious. These techniques are practical in delivery and the therapist makes use of the standard tools of modelling, role-play, feedback, videorecording, etc. For those who are most incapacitated, therapy may be individual initially but ultimately progress to a group setting is necessary to achieve an optimum environment in which to further develop and test the skills being acquired. As the particular traits which respond to these techniques are found in many categories of abnormal personality they may be applied liberally but in particular to those with avoidant or passive-aggressive personality disorder where this conglomerate of traits is central. The benefits of both these techniques have been described by Perry and Flannery (1982) although there have been no large-scale studies.

The use of cognitive therapy alone or combined with behavioural therapy for low self-esteem have found ready advocates in Alden and Safran (1978) and in Beck *et al.* (1990). As with all cognitive approaches the underlying assumptions, called schemata, are identified and the faulty beliefs and behaviours which accrue from these are targeted for change using the standard strategies of cognitive therapy. However, specific attention is also paid to helping patients recognize their maladaptive schemata and then bringing about a shift in these. This is usually a lengthy process so that cognitive behavioural therapy for personality disorders takes far longer than when this approach is used for specific psychiatric disorders, such as depression and anxiety. Unfortunately, so far there has been little evaluation of this approach (Beck *et al.* 1990, pp. 11–21).

Anger and aggression

The appropriate control and expression of anger is paramount for peaceful and amicable social coexistence. Anger may be excessively expressed or controlled. The former is more commonly recognized than the latter and passive-aggressive personality disorder, believed to be a manifestation of the over-control of anger, will not be included in ICD 10 although it is listed in DSM-III-R. As with other abnormal constellations of traits those relating to aggression may be found in many categories but especially in the anti-social personality.

One of the difficulties with therapy is the apparent willingness of patients with anger control problems to enter treatment followed by either sabotage of therapy or alienation of the therapist by aggressive outbursts or acting-out behaviour. Sabotage is a problem for those who are over-controlled and unable to express their feelings and may manifest itself as being repeatedly late for appointments, forgetting to come, etc. For this reason there is little optimism about the psychodynamic therapies for 'anger expression' disorders. However, the application of behavioural techniques, especially assertive training, to assist in the appropriate expression of anger, have received some plaudits (Perry and Flannery 1982) based on individual case studies.

Anger management is a further approach that may be useful for those who are prone to aggressive outbursts. Difficulties with the appropriate expression of anger may also lead to inwardly directed hostility and suicidal gestures. The approach is behavioural with therapy being underpinned by identifying triggers which provoke this behaviour. The patient may be required to re-enact the scene of some prior episode in order to aid this process, and using videotapes and role-play more appropriate behaviours are taught (Maiuro 1987). There have been claims that this approach may be useful in treating men who batter their wives as well as for the more generally violent person (Rosenbaum and Maiuro 1989). If studies examining the hypothesis that sexually assaultive men displace their anger on to their victims prove this to be correct, the approach may in theory be appropriately applied to sexual offenders (Mauiro and Hall 1986). However, this work is still gestational.

As many of those with difficulties in the control and expression of anger are also lacking more general communication skills the addition of social skills may facilitate further improvement in functioning.

Dependence

Those who show traits of excessive dependency span the main categories of abnormal personality, such as antisocial, hysterical, dependent, anxious, etc., and may be treated by exposure to situations requiring autonomous decision-making in a hierarchy, coupled with anxiety management (Turkat and Carlson 1984). Assertiveness and social skills training as well as cognitive therapy are also recommended. While involved in therapy the therapist must be cautioned against being over directive and preventing the development of autonomy in the patient. In particular the use of 'guided discovery' and a Socratic approach to questioning to aid the patient in autonomous decision-making are advocated (Beck *et al.* 1990). However, these approaches to therapy are still anecdotal and need scientific evaluation.

Control, rigidity, and obsessionality

This constellation of traits is found most commonly in those with ananksatic (obsessional) personality disorder but may also be seen in patients with anxious, with dependent, and at times those with antisocial personality. The literature, although scanty, is generally concerned with anankastic personality disorder and a combined psychodynamic and behavioural approach has been recommended anecdotally—the former to aid the patient in dealing with risk and uncertainty, the latter to deal with specific symptoms (APA 1989). These patients generally find group therapy too changeable and unpredictable to complete. More recently, problem-solving has been advocated to deal with procrastination and equivocation, whereas a weekly activity schedule can assist the patient on an hour-by-hour basis. Worrying and anxiety can be relieved by relaxation techniques, and graded task assignments, where tasks are broken down into smaller components, may demonstrate the effectiveness of doing tasks progressively rather than in their entirety. These cognitive and behavioural strategies are outlined clearly by Beck *et al.* (1990).

Flamboyance and eccentricity

These clusters apply to those belonging to the narcissistic, borderline, hysterical or schizotypal categories of personality disorder. The status of some is uncertain and controversial. Borderline personality disorder is perhaps the most controversial but also the most often written about in recent years. From the therapeutic point of view, patients with borderline personality disorder are regarded as difficult and there have been few controlled studies of the impact of intervention. Most of the reports of treatment have been based on case studies, mainly of the application of the various psychotherapeutic approaches (Waldinger 1987; Pines 1989), and in general have been outside the availability of health service patients. Both strict psychoanalytic psychotherapy and the more general psychodynamic approaches are recommended and each has its adherents and critics. On the other hand, some therapists feel that treatment should be strictly structured and supportive and avoid in-depth therapies because of the risk of regression or psychosis. The question of whether therapy should be conducted in an in- or out-patient setting is also unresolved. The Menninger Clinic project (Kernberg *et al.* 1972) did attempt to compare various psychological treatments for this disorder and found that 'expressive psychotherapy' proved superior to either supportive or analytical psychotherapy. However, when reading these studies one is struck by the time and effort involved in treatment and this may constrain the clinician in everyday practice. Recent reports challenge this latter assertion and in particular cognitive analytic therapy coupled with sequential diagrammatic reformulation (Beard *et al.* 1990) has been suggested as one approach to treatment. Essentially, this consists of mapping the shifts in mental state and the mechanisms underlying these changes. Coping modes are identified and the 'core' state of, for example, loneliness or desire to please, identified. A cognitive approach is then applied to bring about the desired change in coping mechanisms. Therapy is time limited and is thus believed to be within the scope of current resources. This approach is also recommended for the narcissistic, hysterical, and antisocial personality.

In general, however, there is a consistent view that borderline, narcissistic, schizotypal, and hysterical personality disorders are difficult to treat and have a guarded prognosis. Individual therapy is the backbone of treatment (Kernberg 1976; Kohut 1977) and group therapy is not recommended because the nature of the difficulties is likely to lead to disruption in the group setting.

Aloofness, suspicion, and sensitivity

These abnormal traits are to be found in extreme degree in those with schizoid, paranoid, schizotypal, and antisocial personality disorders. There is a consensus that patients with this constellation of traits are probably not amenable to psychological treatment because the trust which is essential to the therapeutic relationship is absent. However, there have been individual reports of success using psychodynamic techniques.

Disorder-specific therapies

Therapeutic communities

These are few in number. The treatment of antisocial personality has been studied with greater intensity than any of the other categories, with the possible exception of borderline personality. The role of the therapeutic community in treating the psychopath has received accolades (Whiteley 1970). Treatment includes group as well as individual therapy, and a structured environment designed to increase the patients' responsibility for their actions. Studies to date have shown a decrease in recidivism and symptomatology, and an improvement in social skills (Gunn *et al.* 1978). What is unclear from these studies is the extent to which progress is maintained outside the therapeutic community setting. Family support is also recommended to assist in changing faulty communication patterns (Parsons and Alexander 1973). An adjunct to management is the utilization of crisis intervention techniques at times of crisis although there is no evidence to suggest that this has any long-term impact.

'Wilderness' programmes

An unusual approach to the treatment of offenders diagnosed as being sociopaths is the use of 'wilderness' programmes which have been developed in the USA. These consist of strenuous physical and mental exercises which are conducted in camping and hiking expeditions with other seriously disturbed individuals. The focus during the exercise is on social and interpersonal problems. These programmes are similar in approach to the more protracted therapeutic community programmes. Lasting benefits have been hailed by some (Kimball 1979) although the technique has not migrated across the Atlantic.

Other approaches

In general, operant and aversive techniques are not helpful in treating the psychopath as the effects are temporary and do not generalize. It is uncertain if this absence of success is due to lack of commitment by the therapists or to real therapeutic failure (Moss and Rick 1981). Also, psychodynamic psychotherapy is not successful in this group of patients.

Clinical guidelines

The treatment of personality disorders is in a parlous state and should be addressed as a matter of urgency in a planned and coherent manner, not least in order to reduce the relapse rate from all axis 1 mental state disorders. Moreover it is blatantly obvious that there is a vacuum in relation to the objective appraisal of the treatability of those with personality disorders. Approaches which are time-consuming or over-inclusive are unlikely to be practical and some 'lateral thinking' is necessary if new approaches to intervention are to come to fruition.

1. The first requirement in assessing any therapy is careful delineation of the population under investigation. Even with the advent of DSM-III-R and the use of operational definitions, the process is incomplete and has its critics (Tyrer 1988b). The availability of the tools to assess personality which are independent of axis 1 diagnoses and free from the bias of clinical appraisal will greatly facilitate the important first step. In a clinical setting access to patients for this type of research is likely to be through standard out-patient clinics. As many of these patients will have axis 1 disorders it is essential that detailed personality assessment and therapy be delayed until concomitant axis 1 disorders have resolved.

2. The author would plead that therapy should focus on specific constellations of traits and behaviours such as those outlined in this chapter rather than applying treatment to already unsatisfactory categories of personality disorder. An alternative is to focus on specific cognitive styles and develop therapies around these. This approach is still in its infancy although it has much intuitive appeal.

3. The aim of therapy should be improvement in specific areas of functioning rather than global change. The possibility of effecting total change in the underlying defence mechanisms/cognitions/behaviours may be impractical, if not impossible, in the hurly-burly of everyday clinical work. Thus, the outcome goals should be limited and clearly defined. These goals might include changes in social functioning, quality of life, etc.

4. As behavioural techniques are both easily learned and applied and are more likely to engage the reluctant patient than the cumbersome psychodynamic techniques, there is potential for the clinical use and objective assessment of such methods. The addition of cognitive techniques has intuitive appeal and deserves further assessment. These approaches can be applied in clinical as well as in investigative practice.

It is to be hoped that studies of the impact of treatments, as recommended above, will forthwith be conducted with the rigour that has been applied to pharmacological treatments in general and free from the baggage of anecdote, impressionism, and grandiosity, which to date has been the hallmark and limitation of the treatments for personality disorders.

References

Alden, L. and Safran, J. (1978). Irrational beliefs and non-assertive behaviour. *Cognitive Therapy and Research*, **2**, 357–4.

APA (American Psychiatric Association) (1989). *Treatment of personality disorders*, Vol. 3. American Psychiatric Association, Washington, DC.

Beard, H., Marlowe, M., and Ryle, A. (1990). Management and treatment of personality-disordered patients. The use of sequential diagrammatic reformulation. *British Journal of Psychiatry*, **156**, 541–5.

Beck, A. T., *et al.* (1990). *Cognitive therapy of personality disorders*. Guilford, New York.

Casey, P. and Butler, E. (in prep.). The effect of personality disorder on response to ECT in severe depression.

Gunn, J., Robertson, G., Dell, S., and Way, C. (1978). *Psychiatric aspects of imprisonment*. Academic Press, London.

Kernberg, O. F. (1976). *Object relations theory and clinical psychoanalysis*. Jason Aronson, New York.

Kimball, R. O. (1979). *Wilderness experience program: Final evaluation report*. Health and Environmental Department, Sante Fé, New Mexico.

Kohut, H. (1977). *The restoration of the self*. International Universities Press, New York.

Maiuro, R. D. (1987). Helping angry and violent people manage their emotions and behaviour. *Hospital and Community Psychiatry*, **38**, 1207–10.

Maiuro, R. D. and Hall, G. C. N. (1986). Anger and hostility in sexually assaultive males. *Sexual Coersion and Assault*, **1**, 119–24.

Millon, T. (1981). *Disorders of personality, DSM-III: Axis II*. Wiley, New York.

Moss, G. R. and Rick, G. R. (1981). Overview: applications of operant technology to behaviour disorders in adolescents. *American Journal of Psychiatry*, **138**, 1161–9.

Parsons, B. V. and Alexander, J. F. (1973). Short-term family intervention: a therapy outcome study. *Journal of Consulting Clinical Psychology*, **41**, 195–201.

Perry, J. C. and Flannery, R. B. (1982). Passive-aggressive personality disorder: treatment implications of a clinical typology. *Journal of Nervous and Mental Disease*, **170**, 164–73.

Pines, M. (1989). Borderline personality disorder and its treatment. *Current Opinion in Psychiatry*, **2**, 362–7.

Rosenbaum, A. and Maiuro, R. D. (1989). Eclectic approaches in working with men who batter. In *Treating men who batter: theory and practice* (ed. P. L. Caesar and L. K. Hamberger), pp. 165–95. Springer, New York.

Trower, P., Bryant, B., and Argyle, M. (eds) (1978). *Social skills and mental health*. Pittsburgh University Press, Pittsburgh.

Turkat, I. D. and Carlson, C. R. (1984). Data based versus symptomatic formulation of treatment: the case of a dependent personality. *Journal of Behaviour Therapy and Experimental Psychology*, **15**, 153-60.

Tyrer, P. (ed.) (1988*a*). *Personality disorders: diagnosis, management and course.* Wright, London.

Tyrer, P. (1988*b*). What's wrong with DSM 111 personality disorders? *Journal of Personality Disorders*, **2**, 231-91.

Tyrer, P. and Sievewright, H. (1988). Studies of outcome. In *Personality disorders: diagnosis, management and course* (ed. P. Tyrer), pp. 119-36. Wright, London.

Tyrer, P., Casey, P. R., and Gall, J. (1983). The relationship between neurosis and personality disorder. *British Journal of Psychiatry*, **142**, 404-8.

Waldinger, R. J. (1987). Intensive psychodynamic therapy with borderline patients: an overview. *American Journal of Psychiatry*, **144**, 2267-74.

Whiteley, J. S. (1970). The response of psychopaths to a therapeutic community. *British Journal of Psychiatry*, **116**, 517-29.

Zimmerman, M., Coryell, W., Pfohl, B., Corenthal, C., and Stangl, D. (1986). ECT response in depressed patients with and without a DSM-111 personality disorder. *American Journal of Psychiatry*, **143**, 1030-2.

12

Opiate misusers: are treatments effective?

PHILIP ROBSON

Introduction

The world-wide epidemic of HIV has transformed the political and clinical response to injecting drug use, because it is clearly established as a major factor in the spread of the virus (WHO 1989). Transmission to heterosexual non-injecting partners (France *et al*. 1988), and vertical transmission are of particular concern: in a study of paediatric cases of AIDS, the mothers of 47 per cent of the affected children had injected drugs (Mok 1989). A lack of drug services can produce frightening consequences. In Edinburgh, for example, a culture developed among drug users in which the sharing of injecting equipment was commonplace: the HIV positivity rate in a sample of these people was a horrifying 52 per cent (Brettle *et al*. 1987). Of this number, 20 per cent had shared with users in other parts of the country. This disaster has sombre implications for the population at large: in this same city, screening of a sample drawn from the general practice aged between 20 and 30 years revealed an HIV positivity rate of 1 in 14 men and 1 in 28 women (Robertson 1991, personal communication).

An influential report from the Advisory Council on the Misuse of Drugs —ACMD (1988) has stressed the need to make contact with and offer flexible services to as wide a range of drug misusers as possible, including those whose present motivation is to continue using drugs for pleasure. The Report stresses that 'the spread of HIV is a greater danger to individual and public health than drug misuse'. In particular, it is essential to find ways of inducing users to stop sharing injection equipment, which has been the main factor in transmitting HIV in the addict population, and thence to their partners and offspring (Des Jarlais *et al*. 1985). The impetus to pursue this policy is enhanced by the evidence that HIV education campaigns seem capable of producing lasting changes in the behaviour of illicit drug users in rates of injecting, sharing, and adopting safer sexual practices (Skidmore *et al*. 1989).

In The Netherlands, a rather progressive policy has been adopted for some years with apparent success. The Dutch have recognized that, although all drugs can have harmful primary effects, their illegality creates secondary physical and social problems for both the user and society (Engelsman 1991). Dutch policy has been reviewed in some detail by Engelsman (1989). Some important elements of this policy are as follows: the drug abuse problem is tackled primarily in the context of health and social well-being rather than police and justice; distinctions are made in law between 'drugs with unacceptable risks' (such as opiates and stimulants) and cannabis, which is effectively de-criminalized; attempts are made to normalize and 'demythologize' drug use; prescribed methadone and treatment programmes are widely available; there is awareness of double standards in relation to alcohol and tobacco; and there is willingness to debate the merits of legalisation at some point in the future. It is too early to say with any confidence how successful or otherwise this policy is proving, but the preliminary reports seem encouraging (Engelsman 1989, 1991).

Drug misuse is widespread in our society and numbers dictate that addictions cannot be left to the specialist services, which remain poorly developed in some districts. For decades, general adult psychiatrists have had a central role in drug service planning and delivery, occupying a position of influence which has, at times, been deeply resented by other professional groups and interested parties. It has been remarked that '. . . hospital-based shrinks . . . have a good deal more to learn than teach about drug "misusers" and their motivations' (Tyler 1988). It is certainly true that training in the management of addictive behaviour is unimpressive in some medical schools and postgraduate settings.

However, I believe that the general psychiatrist is well qualified to manage addictive behaviour. Recent neurobiological research (Wise 1988), which raises the possibility of important pharmacological advances, the high incidence of physical problems amongst drug misusers, and the frequent need for constructive and responsible prescribing, all point to the usefulness of a general medical training. Psychiatrists are familiar with a number of psychological techniques, are used to providing long-term support and help with problem-solving, and can cope with emotional people at times of crisis. Chemical addictions share common ground with disorders that are familiar to psychiatric clinics, such as obsessive compulsive disorder or bulimia (Marks 1990): an urge to engage in counter-productive behaviour; mounting tension until sequence completed; temporary reduction of tension and repeat of sequence; external cues specific to particular syndromes; secondary conditioning of urge to external and internal cues; and habituation of craving and withdrawal by cue exposure.

Prevalence of opiate use in the UK is difficult to estimate, but it is widely accepted to be greatly in excess of the figures obtained through notification by doctors to the Home Office and, to rectify this uncertainty, a nationwide database has been established. Non-addictive use of opiates (chipping) by people not in contact with any services is probably not at all uncommon

(Zinberg 1979; Blackwell 1983). One of the reasons given for being reluctant to get involved in trying to help opiate addicts is the feeling that it is a futile task. In fact, it is quite clear that, although many other factors are important in determining outcome, treatment of drug misuse is effective, and that the degree of improvement in terms of drug consumption, medical problems, crime and reliance on illegal income, family problems, and psychological well-being is proportional to the amount of treatment received (McLellan *et al*. 1982).

In the remainder of this chapter I shall be outlining some recent literature on detoxification techniques, substitute prescribing, the effectiveness of counselling and other psychological treatments, psychiatric co-morbidity in drug users, and treatment outcome. I will end with some practical suggestions regarding assessment and management of opiate misusers.

Such a brief chapter is necessarily selective, and the interested reader is referred to Ghodse (1989) for a useful textbook, and Tyler (1988) for a well-informed and controversial journalistic view. Burroughs (1953) has written an extraordinary first hand account of opiate addiction. For a full account of the pharmacology of opioid agonists and antagonists, see Jaffe and Martin (1982).

Detoxification

Most detoxifications occur in the community, either conducted by the subject without recourse to any services, or supported by general practitioners (GPs) or others with symptomatic prescribing or methadone reduction programmes over days, weeks or months. However, clinical experience suggests that such attempts are often unsuccessful, and this impression is supported by a much-quoted study (Gossop *et al*. 1986). An 8 week methadone reduction schedule achieved withdrawal from opiates in only 17 per cent of patients, compared with an 81 per cent success rate for a 3 week in-patient programme in matched subjects randomly allocated to the two treatments. This finding underlines the desirability of having access to detoxification beds in each health district.

Most in-patient units transfer patients on to methadone by titration against withdrawal effects over a 24 hour period, then reduce the methadone in an exponential pattern (so that the reduction remains a fixed proportion of each day's dose). It is useful to have an objective and reliable means of quantifying the severity of the withdrawal syndrome, such as that described by Bradley *et al*. (1987). A typical regime lasts for 21 days, and peak withdrawal effects are to be expected towards the end of this period, persisting for 40 days after abstinence has been achieved or longer. The starting dose of methadone appears to have little effect in determining the severity of the syndrome (Gossop *et al*. 1987*a*).

Shorter regimes, for example 10 days, may result in higher peak withdrawal scores and higher drop-out rates (Gossop *et al.* 1989*b*). Providing clear information to the patient before detoxification begins as to what symptoms might be expected and when may reduce drop-out rates.

Recently, the combination of clonidine and naltrexone has been described as a speedier alternative to methadone reduction in both in-patient (Charney *et al.* 1986; Brewer *et al.* 1988) and out-patient (Kleber *et al.* 1985, 1987) settings. Clonidine reduces withdrawal symptoms by stimulating presynaptic α_2-receptors in locus ceruleus, and probably by activating postsynaptic α_2-receptors in the amygdala, forebrain, and spinal cord. Naltrexone is a long-acting, orally active μ-receptor antagonist which may hasten methadone elimination by displacing it from receptor sites, but the mode of action remains uncertain. Benzodiazepines and other symptomatic treatments are usually given as required. There is some evidence that the technique may be associated with superior outcomes (Milby 1988), but it is not without its critics (Drummond and Turkington 1989).

Prescribing issues

The question of substitute prescribing of opiates to addicts has generated an acrimonious debate extending over several decades. A complete spectrum of views has been aired, ranging from those who argue against any form of substitute prescribing as collusive with dependency (e.g., Rathod 1987) to those who advocate that drugs, including heroin for injection, smoking or snorting, should be readily available to addicts through drug dependency clinics, because prohibition has never worked and merely promotes gangsterism (Marks 1987).

The HIV epidemic has highlighted the need for a 'harm minimization' policy (ACMD 1988), and the current consensus view would be that availability of substitute prescribing forms an essential part of any service for drug users, but that if employed indiscriminately it may prove counterproductive (Strang *et al.* 1987*a*).

A famous clinical trial (Hartnoll *et al.* 1980) illustrates that, when discussing which drug should be used for substitute prescribing, it is not a matter of which is more 'effective', but rather a philosophical decision as to which outcome is more culturally acceptable. In this experiment, intravenous heroin was compared with oral methadone in a study of, for this field, unusual methodological rigour. The patients, who were established opiate addicts adamant in their request for a heroin prescription, were randomly allocated to the two groups and followed-up for a year by independent assessors regardless of whether or not they remained in contact with services. Of the heroin group, 74 per cent remained in treatment for the year of study but, of course, all were continuing to inject; 12 per cent of this group were regularly selling a proportion of their script on to the black market. In the methadone group, 12 per cent broke contact with services immediately on being informed that heroin was not available, and by 12 months only 29 per cent of the original sample remained in contact with services. However, 40 per cent of those not in contact with services stopped using illegal drugs regularly. Overall, the two groups were similar at the end of the study in their average rates of illicit drug use and time spent in the drug culture, unemployment, criminality, and poor general health, but this equivalence conceals a polarization effect in the methadone group: some did unusually well in terms of illicit drug use, crime, and so forth, but a similar proportion fared very badly. It should be noted that the prescribing of heroin to addicts requires a special licence from the Home Office, although other opiates can be administered in injectable form without such a licence. However, many would argue that injectable prescribing has only a very limited role (Battersby *et al.* in press).

Methadone is used in many countries as the substitute opiate of choice, and its pharmacology has been well described (e.g., Kreek 1979). In the UK, it is usually dispensed for oral use as a 1 mg/ml mixture which addicts do not usually inject because of the 'dirty hit' (i.e. unpleasant side-effects) it produces and the inconveniently dilute formulation, but it has some black market value for oral use. There is now a huge literature relating to methadone treatment, much of which emanates from the US, which generally reports favourably on its usefulness. Its critics, who not infrequently argue from the perspective of the clinical psychologist, tend to suggest that its use represents social control rather than treatment and that it is more difficult to give up than heroin (George 1990), or focus on the imperfect methodology of most of the studies (Gossop 1978).

Much of the literature reports on the use of methadone in a long-term, maintenance context and, although it is true that many studies have unsatisfactory selection criteria or lack control groups, the numbers of patients studied are often very large. Retention rates seem to be better than in drug-free programmes (Bass and Brown 1973; Joe and Simpson 1975), which, although not itself a marker of success (Gossop 1978), may be important given the finding that the length of time spent in treatment, regardless of type, is the best predictor of favourable outcome (Simpson 1979; Hser *et al.* 1988). Other studies have reported significant reductions in illicit drug use, criminality, needle-sharing, drug-dealing, unemployment, poor family relationships, and general health (e.g., Ball *et al.* 1988; Hser *et al.* 1988; Paxton *et al.* 1978). When failure of funding caused termination of a methadone programme, major adverse consequences included increases in illicit drug use, drug-dealing, and general criminality (Anglin *et al.* 1989). To achieve best results, the methadone dose must be adequate (Ball *et al.*

1988). Methadone appears to be safe when given over long periods (Kleber 1989).

Psychological interventions

The literature on psychological interventions applied to opiate misusers is small compared to that relating to alcohol, but generally indicates a beneficial effect. Training in recreational and personal skills, with psychotherapy and crisis intervention as indicated, improves outcome in patients maintained on methadone (Stark 1989). Counselling can bring about reductions in prescribed methadone and illicit drug use, unemployment, and arrests, and the extent of the benefit is related to the skill of the individual counsellor (McLellan *et al.* 1988). There is some uncertainty as to whether more specialized psychological treatments can improve outcome. One study suggested that both cognitive-behavioural therapy and 'supportive-expressive psychotherapy' were superior to counselling alone (Woody *et al.* 1983), but this may be related to the time spent with subjects—35 per cent more in the two specialist conditions. Interpersonal psychotherapy showed no advantages over brief, monthly contacts in a six month study of patients on methadone maintenance, although both groups improved (Rounsaville *et al.* 1983). Behavioural methods seem to be effective, at least in the short term: for example, cue exposure with response prevention results in a reduction in drug craving (Bradley and Mooney 1988) but it is not clear if this persists or generalizes beyond the experimental setting.

Two other techniques are worthy of mention because of the breadth of their application, although clear evidence of their efficacy in opiate addicts is still awaited. 'Motivational interviewing' (Miller 1983) takes as its starting point an attitude toward the patient and his or her difficulties which is different from the traditional view. Individuals are expected to take full responsibility for their actions and their consequences, encouraged in the belief that they have the power to shape their own destiny, and helped in developing the skills and confidence needed to achieve personal goals. 'Denial' is seen not as a ubiquitous and immovable character trait, but rather as a product of a confrontative client/therapist interaction without genuine rapport. The therapist aims, by a combination of reflective listening, Socratic dialogue, and various cognitive-behavioural techniques, to induce a state of cognitive dissonance between the client's self-view and the maladaptive behaviour. Resolution of the dissonance by a change of behaviour is then encouraged by a focus on improving self-esteem and self-efficacy.

'Relapse prevention' owes much to the work of Marlatt and colleagues (e.g., Marlatt and George 1984). Here, the aim is to plan ahead in order to foresee 'high-risk situations' (HRS) and the subtle personal decisions and life-style factors which may lead to them: for example, an abstinent alcoholic who decides to keep a bottle in the house just in case an old drinking buddy might call round. HRSs are either avoided or confronted with rehearsed coping responses which, if successful, lead to an increased sense of self-efficacy and reduced probability of future relapse. If, on the other hand, they fail, self-efficacy is undermined and positive memories of drug-effects may be triggered. A single lapse may induce the abstinence violation effect—'that's it, I've had a hit, all that effort's been a complete waste of time, I'll always be a junky, might as well go the whole hog'—and lapse becomes relapse.

Psychiatric co-morbidity

Psychiatric symptoms occur very frequently in opiate misusers (with General Health Questionnaire 'caseness' as high as 61 per cent (Swift *et al.* 1990)) and, if of high intensity, are associated with poor outcome (McLellan *et al.* 1983). Depressive symptoms and anxiety commonly occur (Rounsaville *et al.* 1982; Swift *et al.* 1990), but tend to be labile and short-lived, responding to treatment of the addictive behaviour (Kleber *et al.* 1983). The prevalence of major depressive disorder is also increased but this, too, often improves with attention to the drug misuse (Rounsaville *et al.* 1982), and antidepressants should usually be reserved for those with very severe or persistent depression (Kleber *et al.* 1983). Depression can be associated with an increase in illicit drug use (Rounsaville *et al.* 1982), and physicians should bear in mind the possible drug interactions and be aware that prescribed antidepressants may become part of the chaotic drug use.

Many attempts have been made to relate 'personality disorder' to opiate misuse, both as an aetiological factor and as an explanation of unsatisfactory response to treatment. However, these efforts have been fatally flawed by such methodological and conceptual difficulties as reliance on retrospective personality assessment using unreliable measures of unproven validity, use of highly selected subgroups, and confusion of cause with effect.

Treatment outcome and long-term follow-up

It is very clear that detoxification alone has no effect on long-term outcome (e.g., Vaillant 1988), indeed some individuals may undergo hundreds of detoxifications in the course of their addiction career.

When detoxification is linked with some continuing treatment, results are often better than at first seems the case. Although 71 per cent of addicts in one study had used opiates within six weeks of completing detoxification, many of them did not return to dependent use and there was a gradual increase in the number of abstinent users over time, so that at six months

45 per cent were off drugs and living in the community (Gossop *et al.* 1987*b*). Alcohol substitution was not a problem generally, and ability to avoid other drug users was an important prognostic factor. The greatest number of lapses occur in the first week after detoxification and these are often associated with one or more of the following: negative mood states or environmental conditions, unresolved withdrawal symptoms, interpersonal conflicts, reduction of staff support, ready availability of drugs, and return to the drug culture (Bradley *et al.* 1989). Context is highly influential: amongst American soldiers who became addicted to heroin while serving in Vietnam, 56 per cent abstained entirely on their return home and a further 32 per cent who did use it again did not become re-addicted (Robins 1979). Consistent with this is the finding that change of residence after discharge from in-patient or residential care may improve chances of remaining abstinent very considerably (Maddux and Desmond 1982). Time spent in treatment, regardless of type, is proportional to outcome, and patients undergoing treatments less than three months in duration may not differ on long-term follow-up from those receiving no treatment or detoxification only (Simpson 1979). Unfortunately, as many as 50 per cent of patients may leave drug-free residential or out-patient programmes within the first three months (Joe and Simpson 1975).

Three psychological models have been put forward to explain relapse. The conditioned withdrawal model (Wikler and Pescor 1967) suggests that repeated experience of withdrawal symptoms leads to a process of classical conditioning by which environmental cues become conditioned stimuli to aversive reactions in the absence of pharmacological withdrawal. Siegel (1975) has proposed the conditioned tolerance model: here it is argued that the organism attempts to maintain internal homeostasis by compensatory responses opposite to the drug effect, and that these compensatory responses are also conditionable to environmental cues. Finally, Stewart and others (1984) have put forward evidence that conditioned stimuli (and priming doses of drugs acting on particular neurochemical systems) can generate 'positive appetitive states' which mimic the reinforcing effects of the abused drug in its absence, and produce craving unaccompanied by withdrawal symptoms.

Predictions for poor outcomes are very unreliable, but there may be weak associations with high levels of pre-treatment criminality, heavy alcohol use before or during treatment, and living with an addict during treatment (Judson and Goldstein 1982). Recent figures from the US reveal a trend towards higher success rates in achieving abstinence in methadone-maintenance patients, which seems to be related to the advent of briefer detoxification procedures and the availability of psychotherapy (Milby 1988). Matching patients to treatment type must include an assessment of the existence and intensity of psychiatric symptoms (McLellan *et al.* 1983).

A number of attempts have been made to follow the progress of opiate misusers over years rather than months (Vaillant 1988; Stimson *et al.* 1978; Thorley 1981; Cottrell *et al.* 1985; Edwards and Goldie 1987). Opiate addiction is seen to be a chronic relapsing and remitting condition with a significant mortality rate (10–15 per cent at 10 years). Despite this, outcome is not as gloomy as generally thought, because up to 50 per cent of subjects are found to be consistently abstinent at 10 years. The natural history of the condition seems slanted towards recovery (Strang *et al.* 1987*b*), so that the beneficial effect of treatment may be overestimated. Vaillant (1988) suggests certain key pointers to a good outcome: employed for more than half of adult life, and especially for the four years before treatment starts; brought up in parents' own culture; and married at any time. He emphasizes the need to take a long-term view in treatment, underlines the importance of helping patients to find alternate sources of gratification, and points out that, if nothing else, treatment programmes can do a great deal to reduce mortality and suffering.

Clinical guidelines

1. *Support and advise general practitioners*

General practitioners (GPs) have an essential role to play in the management of drug misusers, although many are understandably reluctant to do so. They can improve detection by asking more patients specifically about their use of alcohol and drugs, and by being alert for physical and psychological signs and symptoms of undisclosed use. Guidance should be given by the psychiatrist in accurate assessment, appropriate investigations, and ways to verify the patient's account by clinical observation and recourse to sources of information, such as the Home Office Register, or other clinical teams. Special care must be taken in assessing temporary patients, or those not personally known to the doctor.

The value of counselling and simple problem-solving techniques should be stressed. GPs should be encouraged to prescribe for symptomatic relief in the short term to patients attempting to remain abstinent from opiates and displaying objective withdrawal symptoms. They should also be encouraged to prescribe substitute opiates to patients who wish to undergo detoxification in the community. Methadone mixture (1 mg/ml) is the drug of choice and, after a few days stabilization, the patient and doctor should agree a reduction schedule which can then be formalized in a written contract. Unless there are exceptional circumstances, the drug is best prescribed on a daily pick-up basis, and lost or stolen medication should not be replaced. Random observed urine screens should be arranged and sessions with a drug counsellor organized. Other psychoactive medication is best avoided, support and encouragement being provided liberally in its place.

Table 15 Drugs of misuse: some adverse effects

Drug class	Physical	Psychological
Opiates		
Acute	Drowsiness, ataxia, miosis, respiratory depression, histamine release.	Confusion.
Chronic	Tolerance, hormonal effects, constipation.	Constant preoccupation.
Withdrawal	Yawning, running eyes and nose, tremor, sweating, mydriasis, restlessness, G.I. disturbance, sleeplessness, muscle/bone ache, BP ↑, tachycardia.	Anxiety, irritation, extreme craving.
Stimulants		
Acute	Over-activity, insomnia, sweating, diarrhoea, dysrhythmias, BP ↑ or ↓, hyperpyrexia, convulsions, coma.	Binges/crashes, irritability, panic, delusions, hallucinations.
Chronic	Tolerance, microvascular damage, dopaminergic neuron damage.	Depression, extreme preoccupation, chronic psychosis.
Withdrawal	Deep sleep, lethargy.	Depression, anhedonia, motivation↓, extreme craving.
Barbiturates		
Acute	Drowsiness, ataxia, nystagmus, slurred speech, respiratory depression, BP ↓ , coma.	Confusion, depression, emotional instability, amnesia.
Chronic	As above, selective tolerance, drug interactions.	Aggression, paranoid ideas, psychosis.
Withdrawal	Insomnia, nausea, vomiting, mydriasis, BP ↓ , tremor, muscle tone ↑, reflexes ↑ , convulsions.	Anxiety, irritability, emotional instability, psychosis.
Benzodiazepines		
Acute	Drowsiness, ataxia, vertigo, slurred speech, respiratory depression (massive doses only).	Confusion, amnesia, rebound anxiety.

	Physical effects	Psychological effects
Chronic	Neuroradiological changes.	Cognitive impairment, reduced coping skills.
Withdrawal	Sensory perception ↑, insomnia, vertigo, cramps, G.I. disturbance, convulsions.	Mild to moderate craving, anxiety, depression, paranoid psychosis (rare).
Cannabis		
Acute	Pulse ↑, BP ↓, red eyes, yawning, diuresis, mydriasis, carbohydrate hunger.	Thought disorder, panic, paranoid ideas, psychosis (rare).
Chronic	Bronchitis, accumulation effects, ? ↓ immune system, ? ↓ fertility.	Anxiety, ? motivation ↓, possible psychosis.
Withdrawal	Probably none.	Mild/moderate craving.
Hallucinogens		
Acute	Paraesthesiae, tremor, sympathetic and parasympathetic stimulation, leucocytosis.	Hallucinations, panic, anger, delusions, impulsive behaviour.
Chronic	? Teratogenic, ? neoplastic.	Mood lability, compulsive use, psychosis, suicidal impulses.
Withdrawal	None.	Moderate craving, flashbacks.
Solvents		
Acute	Ataxia, slurred speech, dysrhythmias, convulsions, coma (asphyxiation).	Poor judgement, psychosis, aggression, omnipotent feelings.
Chronic	Hepatic or renal damage, bone marrow depression, neuropathy, encephalopathy, 'glue sniffer's rash'.	As above.
Withdrawal	None.	Moderate craving.
MDMA ('Ecstasy')		
Acute	Cardio-respiratory collapse due to impurities (rare).	Hallucinations (rare).
Chronic	? brain damage (ill defined), ? immune suppression.	Fatigue, depression, binge use.
Withdrawal	None.	Flashbacks.

GPs will be more willing to take on these patients if they are confident that there is a psychiatrist available to give advice on the telephone, and accept referral of patients who do not seem straightforward. After assessment in the out-patient clinic or in the community, it will often prove possible to hand back the care of such patients to the GP.

2. *Enquire about the use of alcohol and drugs*

The response to problems with stimulants, hallucinogens, cannabis, and solvents is symptomatic and supportive. Sedative drugs can be withdrawn gradually according to published guidelines (Medical Working Group on Drug Dependence 1984; Ghodse 1989). If uncertain how to proceed, consult the specialist services. Adverse effects of a number of commonly abused drugs are shown in Table 15.

3. *Attend to the patient's physical health*

As drug users frequently neglect health, a physical examination should be completed, blood taken for a full blood count, serum biochemistry, liver function tests, and screening for hepatitis B. (All those who test negative for the latter should be offered active immunization.) A chest X-ray is desirable. It is essential to provide education concerning HIV and, if the test is requested, to ensure that the patient receives pre- and post-test counselling (Catalan 1990) from someone experienced in this field.

4. *Tailor the response flexibly to individual requirements*

Objective confirmation of the patient's account is desirable where possible, although if genuine rapport can be established, addicts can be surprisingly truthful (Bennett and Wright 1986)! An observed urine specimen should be subjected to a drug screen. Options might include in- or out-patient detoxification, some form of longer-term prescribing, counselling or psychotherapy (individual or group), marital or family therapy, behavioural techniques, or practical help with social or legal difficulties. Referral to a variety of residential rehabilitation units, such as therapeutic communities or Minnesota Method clinics, should be considered if the problems appear deep-seated or if the patient wishes it. Choice of unit will depend upon which concept has credibility for the patient, how long a stay is feasible, and availability of funding. Self-help groups, such as Narcotics Anonymous, are very helpful to some people. Advice from local specialist services should be sought whenever possible in planning management. Ideally, prescribing and opiate-free facilities should exist in separate places within the District, as there is some evidence that each attracts a different population of users (Drummond *et al.* 1986). Needle-exchange should be available and, perhaps surprisingly, does not seem to increase the frequency of illicit IV use (Wolk *et al.* 1990), although this remains a theoretical possibility (Ghodse *et al.* 1987). Many drug users now deny sharing injecting equipment, but the

likelihood is that this continues to correlate closely with frequency of injecting (Ball *et al.* 1988).

5. *Be prepared to prescribe where appropriate*

Symptomatic treatment of opiate withdrawal with diphenoxylate, hyoscine,
clonidine or benzodiazepines may be required. If substitute prescribing of
opiates is indicated, the general psychiatrist is well advised to use methadone mixture 1 mg/ml DTF. Other opiates and injectables are best left to
specialists. Methadone is roughly equipotent with pharmaceutical morphine
and heroin, and 40 mg approximates to 0.5 g street heroin. Some other
approximate equivalents to methadone are as follows: Diconal (dipipanone)
10 mg = 3–5 mg; codeine 30 mg = 1–2 mg; Palfium (dextromoramide) 5 mg
= 5 mg; pethidine 50 mg = 5 mg; buprenorphine 0.2 mg = 2–3 mg; Gee's
Linctus or Dr Collis Brown Mixture 100 ml = 10 mg (Medical Working
Group on Drug Dependence 1984). A better way to arrive at an appropriate
stabilization dose than reliance on equivalence tables is to carry out a
titration against objective withdrawal effects (such as sweating, tremor,
running eyes and nose, yawning, pilo-erection, and vomiting). On day 1,
methadone 20 mg is given and, after a brief observation period, the patient
is asked to return the following morning. If withdrawal symptoms are
present, the dose can be increased by 10 mg and the process repeated until
withdrawal is no longer evident. This procedure can be completed much
more rapidly if a day's admission is possible. In undertaking prescribing,
and indeed any other form of intervention, there must be some very clear
goal in mind so that the value of the treatment can be objectively assessed.
This goal may be modest and be well short of any ultimate goal of abstinence. It is sometimes necessary to take a long view, but maintenance treatment should be seen as a last resort: the aim is to 'coax and cajole along
natural pathways to recovery' (Strang *et al.* 1987*a*). Although it is to be
hoped that, where prescribing is indicated, a steady reduction in dose can be
agreed with the patient, strategic temporary increases can sometimes re-
establish stability when set-backs occur (Gossop *et al.* 1982). It is important
not to try to move too fast in encouraging the process of change (Prochaska
and DiClemente 1983).

6. *Dealing with the pregnant opiate addict*

Random testing in antenatal clinics reveals a 1–2 per cent point prevalence
of unrecognized opiate use (London *et al.* 1990). Opiates taken during
pregnancy may be associated with premature delivery, abruption of the
placenta, and small-for-dates babies (Riley 1987). Fillers in street heroin
and many other drugs of abuse are actively teratogenic. On the other hand,
rapid withdrawal of opiates or intermittent use can lead to intra-uterine
death. Mothers may be undernourished and vitamin-deficient, anaemic,
and prone to infections and depression.

Careful liaison with the GP and obstetrician is essential. Substitute prescribing is often necessary, and the aim is to identify the lowest dose of methadone that will effectively prevent withdrawal and recourse to other drugs or alcohol. A very gradual reduction is then undertaken with the optimal aim of becoming drug-free 6 to 8 weeks before expected delivery. HIV testing is advisable because of the high possibility of vertical transmission and the possible worsening of prognosis in the mother.

Some withdrawal symptoms, such as irritability, twitching, and sneezing in the neonate are likely if the maternal methadone dose is 15 mg or more (Riley 1987), but respiratory depression is rare. Mild symptoms can be treated with simple supportive measures, such as soft lights, quiet environment, and swaddling. If more troublesome, morphine sulphate solution (5 mg/10 ml) 0.2 ml can be given 3 hourly, increasing by 0.05 ml until relief is obtained. This can be withdrawn at the rate of 0.05 ml on alternate days (Kandall *et al.* 1983). Chlorpromazine can be used as an alternative, but benzodiazepines must be avoided as they paradoxically lower the convulsion threshold in these patients.

7. Opiate addicts requiring surgery or pain relief

The addict's established daily dose of opiate should be regarded as a physiological requirement, with pre-medication and post-operative analgesia being considered quite separately (Wood and Soni 1989). Post-operatively, this requirement should be satisfied by the use of intramuscular methadone until oral medication can be resumed.

General anaesthesia may produce unexpected hypotension in narcotic users, and local techniques are preferable where feasible. Post-operative analgesia should be supplied as would normally be appropriate for the given procedure, but larger doses may be required if there is objective evidence of continuing pain. Agonist/antagonist analgesics, such as pentazocine or buprenorphine, may precipitate a withdrawal syndrome.

References

(ACMD) Advisory Council on Misuse of Drugs (1988). *AIDS and drug misuse.* HMSO, London.

Anglin, M. D., Speckart, G. R., Booth, M. W., and Ryan, T. R. (1989). Consequences and costs of shutting off methadone. *Addictive Behaviours*, **14**, 307–26.

Ball, J. C., Lange, W. R., Myers, C. P., and Friedman, S. R. (1988). Reducing the risk of AIDS through methadone maintenance treatment. *Journal of Health and Social Behaviour*, **29**, 214–26.

Bass, U. F. and Brown, B. S. (1973). Methadone maintenance and methadone detoxifications: a comparison of retention rates and client characteristics. *International Journal of Addiction*, **8**, 889–95.

Battersby, M., Farrell, M., Gossop, M., Robson, P., Strang, J. (in press). 'Horse trading', prescribing injectable opiates to opiate addicts—a descriptive study. *British Journal of Addiction*,

Bennett, T. and Wright, R. (1986). The impact of prescribing on the crimes of opioid users. *British Journal of Addiction*, **81**, 265–73.

Blackwell, J. S. (1983). Drifting, controlling, and overcoming: opiate users who avoid becoming chronically dependent. *Journal of Drug Issues*, **13**, 219–35.

Bradley, B. P., Gossop, M., Phillips, G. T., and Legarda, J. J. (1987). The development of an opiate withdrawal scale (OWS). *British Journal of Addiction*, **82**, 1139–42.

Bradley, B. P. and Mooney, S. (1988). Extinction of craving during exposure to drug-related cues: Three single case reports. *Behavioural Psychotherapy*, **16**, 45–56.

Bradley, B. P., Phillips, G., Green, L., and Gossop, M. (1989). Circumstances surrounding the initial lapse to opiate use following detoxification. *British Journal of Psychiatry*, **154**, 354–9.

Brettle, R. P., *et al.* (1987). Human immunodeficiency virus and drug misuse: the Edinburgh experience. *British Medical Journal*, **295**, 421–4.

Brewer, C., Rezae, H., and Bailey, C. (1988). Opioid withdrawal and naltrexone induction in 48–72 hours with minimal dropout, using a modification of the naltrexone-clonidine technique. *British Journal of Psychiatry*, **153**, 340–3.

Burroughs, N. (1953). *Junky*. Penguin, London.

Catalan, J. (1990). HIV and AIDS-related psychiatric disorder: what can the psychiatrist do? In *Dilemmas and difficulties in the management of psychiatric patients* (ed. K. Hawton and P. Cowen), pp. 205–17. Oxford University Press.

Charney, D. S., Heringer, G. R., and Kleber, H. D. (1986). Combined use of clonidine and naltrexone as a rapid, safe and effective treatment of abrupt withdrawal from methadone. *American Journal of Psychiatry*, **143**, 831–7.

Cottrell, D., Childs-Clarke, A., and Ghodse, A. H. (1985). British opiate addicts: an 11-year follow-up. *British Journal of Psychiatry*, **146**, 448–50.

Des Jailais, D. C., Friedman, S. R., and Hopkins, W. (1985). Risk reduction for the acquired immune deficiency syndrome among I.V. drug users. *Annals of Internal Medicine*, **103**, 755–9.

Drummond, C. B. and Turkington, D. (1989). Naltrexone and clonidine in heroin withdrawal treatment. *British Journal of Psychiatry*, **154**, 571–2.

Drummond, C. B., Colin, D., Taylor, J. A., and Mullin, P. J. (1986). Replacement of a prescribing service by an opiate-free day programme in a Glasgow drug clinic. *British Journal of Addiction*, **81**, 559–65.

Edwards, J. G. and Goldie, A. (1987). A ten-year follow-up study of Southampton opiate addicts. *British Journal of Psychiatry*, **151**, 679–83.

Engelsman, E. L. (1989). Dutch policy on the management of drug-related problems. *British Journal of Addiction*, **84**, 211–18.

Engelsman, E. L. (1991). Drug misuse and the Dutch. *British Medical Journal*, **302**, 484–5.

France, A. J., *et al.* (1988). Heterosexual spread of human immunodeficiency virus in Edinburgh. *British Medical Journal*, **296**, 526–9.

George, M. (1990). Methadone screws you up. *The International Journal on Drug Policy*, **1**(5), 24–5.

Ghodse, H. (1989). *Drugs and addictive behaviour—a guide to treatment.* Blackwell, Oxford.

Ghodse, A. H., Tregenza, G., and Li, M. (1987). Effect of fear of AIDS on sharing injection equipment among drug abusers. *British Medical Journal*, **295**, 698–9.

Gossop, M. (1978). A review of the evidence for methadone maintenance as a treatment for narcotic addiction. *Lancet*, **1**, 812–15.

Gossop, M., Strang, J., and Connell, P. H. (1982). The response of out-patient opiate addicts to the provision of a temporary increase in their prescribed drugs. *British Journal of Psychiatry*, **141**, 338–43.

Gossop, M., Johns, A., and Green, L. (1986). Opiate withdrawal: inpatient vs outpatient programmes and preferred vs random assignment to treatment. *British Medical Journal*, **293**, 103–4.

Gossop, M., Bradley, B., and Phillips, G. T. (1987*a*). An investigation of withdrawal symptoms shown by opiate addicts during and subsequent to a 21-day in-patient methadone detoxification procedure. *Addictive Behaviours*, **12**, 1–6.

Gossop, M., Green, L., Phillips, G., and Bradley, B. (1987*b*). What happens to opiate addicts immediately after treatment: prospective follow-up study. *British Medical Journal*, **294**, 1377–80.

Gossop, M., Green, L., Phillips, G., and Bradley, B. (1989*a*). Lapse, relapse and survival among opiate addicts after treatment—a prospective follow-up study. *British Journal of Psychiatry*, **154**, 348–53.

Gossop, M., Griffiths, P., Bradley, B., and Strang, J. (1989*b*). Opiate withdrawal symptoms in response to 10-day and 21-day methadone withdrawal programmes. *British Journal of Psychiatry*, **154**, 360–3.

Hartnoll, L., *et al.* (1980). Evaluation of heroin maintenance in controlled trial. *Archives of General Psychiatry*, **37**, 877–84.

Hser, Y-I., Anglin, M. D., and Chou, C-P (1988). Evaluation of drug abuse treatment—a repeated measures design assessing methadone maintenance. *Evaluation Review*, **12**, 547–70.

Jaffe, J. H. and Martin, W. R. (1982). Opioid analgesics and antagonists. In *The pharmacological basis of therapeutics* (ed. L. S. Goodman and A. Gillman), pp. 491–549. Macmillan: New York.

Joe, G. W. and Simpson, D. D. (1975). Retention in treatment of drug abusers: 1971–72 DARP admissions. *American Journal of Drug and Alcohol Abuse*, **2**, 63–71.

Judson, B. A. and Goldstein, A. (1982). Prediction of long-term outcome for heroin addicts admitted to a methadone maintenance program. *Drug and Alcohol Dependence*, **10**, 383–91.

Kandall, S. R., Doberczak, T. M., Maver, K. R., Strashun, R. H., and Korts, D. C. (1983). Opiate versus central nervous system depressant therapy in neonatal drug abstinence syndrome. *American Journal of Diseases of Childhood*, **137**, 378–82.

Kleber, H. D. (1989). Treatment of drug dependence: what works. *International Review of Psychiatry*, **1**, 81–100.

Kleber, H. D., Weissman, M. M., Rounsaville, B. J., Wilber, C. H., Prusoff, B. A., and Riordan, C. E. (1983). Imipramine as treatment for depression in addicts. *Archives of General Psychiatry*, **40**, 649–53.

Kleber, H. D., *et al.* (1985). Clonidine in outpatient detoxification from methadone maintenance. *Archives of General Psychiatry*, **42**, 391–4.

Kleber, H. D., Topazian, M., Gaspari, J., Riordan, C. E., and Kosten, T. (1987).

Clonidine and naltrexone in the out-patient treatment of heroin withdrawal. *American Journal of Drug-Alcohol Abuse*, **13**, 1–17.

Kreek, M. J. (1979). Methadone in treatment: physiological and pharmacological issues. In *Handbook on drug abuse* (ed. R. L. Dupont, A. Goldstein, and J. O'Donnell), pp. 57–86. NIDA, Washington, DC.

London, M., Caldwell, R., and Lipsedge, M. (1990). Services for pregnant drug users. *Psychiatric Bulletin*, **14**, 12–15.

Maddux, J. F. and Desmond, D. P. (1982). Residence relocation inhibits opioid dependence. *Archives of General Psychiatry*, **39**, 1313–17.

Marks, J. (1987). State-rationed drugs. *Druglink*, **2**, 14.

Marks, I. (1990). Behavioural (non-chemical) addictions. *British Journal of Addiction*, **85**, 1389–94.

Marlatt, G. A. and George, W. H. (1984). Relapse prevention: Introduction and overview of the model. *British Journal of Addiction*, **79**, 261–73.

McLellan, A. T., Luborsky, L., Woody, G. E., O'Brien, C. P., and Druley, K. A. (1982). Is treatment for substance abuse effective? *Journal of the American Medical Association*, **247**, 1423–8.

McLellan, A. T., Luborsky, L., O'Brien, W. G. E., and Druley, K. A. (1983). Predicting response to alcohol and drug abuse treatments. *Archives of General Psychiatry*, **40**, 620–5.

McLellan, A. T., Woody, G. E., Luborsky, L., and Geohl, L. (1988). Is the counsellor an 'active ingredient' in substance abuse rehabilitation? (An examination of treatment success among four counsellors.) *Journal of Nervous and Mental Disease*, **176**, 423–30.

Medical Working Group on Drug Dependence (1984). *Guidelines of good clinical practice in the treatment of drug misuse*. DHSS, London.

Milby, J. B. (1988). Methadone maintenance to abstinence—how many make it? *Journal of Nervous and Mental Disease*, **176**, 409–22.

Miller, W. R. (1983). Motivational interviewing with problem drinkers. *Behavioural Psychotherapy, Clinical Section*, **11**, 147–72.

Mok, J. Y. O. (1989). Vertical transmission of HIV: a prospective study. *Archives of Disease in Childhood*, **64**, 1140–5.

Paxton, R., Mullin, P., and Beattie, J. (1978). The effects of methadone maintenance with opioid takers. (A review and some findings from one British city.) *British Journal of Psychiatry*, **132**, 473–81.

Prochaska, J. O. and DiClemente, C. C. (1983). Stages and processes of self-change of smoking: toward an integrative model of change. *Journal of Consulting and Clinical Psychology*, **51**, 390–5.

Rathod, N. (1987). Substitution is not a solution. *Druglink*, **2**, 16.

Riley, D. (1987). The management of the pregnant drug addict. *Bulletin of the Royal College of Psychiatrists*, **11**, 362–5.

Robins, L. H. (1979). Addict careers. In *Handbook on drug abuse* (ed. R. L. Dupont, A. Goldstein, and J. O'Donnell), pp. 325–36. NIDA, Washington, DC.

Rounsaville, B. J., Weissman, M. M., Crits-Christoph, K., Wilber, C., and Kleber, H. (1982). Diagnosis and symptoms of depression in opiate addicts. *Archives of General Psychiatry*, **39**, 151–6.

Rounsaville, B. J., Glazer, W., Wiber, C. H., Weissman, M. M., and Kleber, H. D. (1983). Short-term interpersonal psychotherapy in methadone-maintained opiate addicts. *Archives of General Psychiatry*, **40**, 629–36.

Siegel, S. (1975). Evidence from rats that morphine tolerance is a learned response. *Journal of Comparative and Physiological Psychology*, **89**, 498–506.

Simpson, D. (1979). The relation of time spent in drug abuse treatment to post-treatment outcome. *American Journal of Psychiatry*, **136**, 1449–53.

Skidmore, C. A., Robertson, J. R., and Roberts, J. J. K. (1989). Changes in HIV risk taking behaviour in intravenous drug users: a second follow-up. *British Journal of Addiction*, **84**, 695–6.

Stark, M. J. (1989). A psycho-educational approach to methadone maintenance treatment. *Journal of Substance Abuse Treatment*, **6**, 169–81.

Stewart, J., de Wit, H., and Eikelboom, R. (1984). Role of unconditioned and conditioned drug effects in the self-administration of opiates and stimulants. *Psychological Review*, **91**, 251–68.

Stimson, G. V., Oppenheimer, E., and Thorley, A. (1978). Seven-year follow-up of heroin addicts: drug use and outcome. *British Medical Journal*, **1**, 1190–2.

Strang, J., Ghodse, H., and Johns, A. (1987a). Responding flexibly but not gullibly to drug addiction. *British Medical Journal*, **295**, 1364.

Strang, J., Heathcote, S., and Watson, P. (1987b). Habit-moderation in injecting drug addicts. *Health Trends*, **19**, 16–18.

Swift, W., Williams, G., Neill, O., and Grenyer, B. (1990). The prevalence of minor psychopathology in opioid users seeking treatment. *British Journal of Addiction*, **85**, 629–34.

Thorley, A. (1981). Longitudinal studies of drug dependence. In *Drug problems in Britain: a review of ten years* (ed. G. Edwards and C. Busch). Academic Press, London.

Tyler, A. (1988). *Street drugs*. Hodder and Stoughton, London.

Vaillant, G. E. (1988). What can long-term follow-up teach us about relapse and prevention of relapse in addiction? *British Journal of Addiction*, **83**, 1147–57.

WHO (World Health Organization) (1989). AIDS surveillance in Europe. *Quarterly Report No. 22*. Collaborative Centre on AIDS, Geneva.

Wikler, A. and Pescor, F. T. (1967). Classical conditioning of a morphine abstinence phenomenon, reinforcement of opioid-drinking behaviour and 'relapse' in morphine addicted rats. *Psychopharmacologia*, **10**, 255–84.

Wise, R. A. (1988). The neurobiology of craving: implications for the understanding and treatment of addiction. *Journal of Abnormal Psychology*, **97**, 118–32.

Wolk, J., Wodak, A., Guinan, J. J., Macaskill, P., and Simpson, J. M. (1990). The effect of a needle and syringe exchange on a methadone maintenance unit. *British Journal of Addiction*, **85**, 1445–50.

Wood, P. R. and Soni, N. (1989). Anaesthesia and substance abuse. *Anaesthesia*, **4**, 672–80.

Woody, G. E., *et al.* (1983). Psychotherapy for opiate addicts—does it help? *Archives of General Psychiatry*, **40**, 639–45.

Zinberg, N. E. (1979). Non-addictive opiate use. In *Handbook on drug abuse* (ed. R. L. Dupont, A. Goldstein, and J. O'Donnell), pp. 303–14. NIDA, Washington, DC.

13

Relapse in schizophrenia: what are the major issues?

EVE C. JOHNSTONE

Introduction

Listing of the major issues regarding relapse in schizophrenia is not difficult, but making decisions about how to deal with them is less easy.

One major issue is the extent to which relapse of some kind is characteristic of the generality of schizophrenic illnesses and another is the type of deterioration which should be included within the definition of relapse. Further important aspects concern possible methods of reducing the number and severity of relapses and the benefits and risks associated with these interventions.

Risk of relapse

The risk of relapse in schizophrenia received substantial study in the years before effective treatments were available. Kraepelin (1919/1971) defined the concept which later came to be known as schizophrenia very largely upon the basis of the poor prognosis of this condition. Essentially, he considered that the development of a defect state which was not completely reversible was the defining characteristic of this disorder, but that the course run to reach this end state was variable, and that although some illnesses took a progressive downhill course, many were characterized by partial remissions and by repeated relapses. He did, however, recognize that a small proportion of patients (13 per cent of his own cases) with this disorder recovered from their psychotic symptoms and remained well for a prolonged period, if not indefinitely.

Others have considered that the proportion who remain well is higher than Kraepelin's estimate, for example, Bleuler (1978) has asserted that 'at least 25% of all schizophrenics recover entirely and remain recovered for good', even though they do not receive neuroleptic medication. Not all, however, would agree with Bleuler's criteria for recovery, which allow the persistence of delusions and perceptual disturbance. The work of Vaillant (1963) describes cases of schizophrenia characterized by remission rather than relapse, but indicates that these are rare and that for most schizophrenic patients the risk of repeated relapse is very real.

The introduction of the widespread use of neuroleptic drugs removed the opportunity of studying the course of the generality of schizophrenia unmodified by antipsychotic agents, but it is clear from the studies described below (e.g., Leff and Wing, 1971; Hogarty *et al.* 1974; Crow *et al.* 1986) that even with neuroleptics more than 40 per cent of patients who enter relevant treatment trials will relapse within two years. Schizophrenic patients who are suitable for and will co-operate with prolonged studies of maintenance treatment are far from representative of the samples from which they are drawn (Johnstone *et al.* 1986; Crow *et al.* 1986), but some idea of the likelihood of relapse in the generality of schizophrenic patients may be obtained from a recent naturalistic follow-up study of all (532) schizophrenic patients discharged from beds providing psychiatric care for patients from the Harrow area of London between 1975 and 1985 (Johnstone *et al.* 1991*a*). The mean number of admissions of these cases was 5.37 (range 1–40), and the mean duration of in-patient care between 1975 and 1985 was 13.7 months. A total of 87 per cent had at least one readmission. Clearly, therefore, even now relapse may be expected to take place in most schizophrenic patients.

The nature of relapse

It has become customary to classify the typical abnormalities of the mental state of schizophrenic patients into positive and negative features with reference to behavioural excesses and deficits. Positive features are pathological by their presence, and negative features represent the loss of some normal function. Positive features are generally considered to include delusions, hallucinations, and positive formal thought disorder (Fish 1962). Some studies (e.g., Johnstone *et al.* 1978) have also included incongruity of affect under this heading. Negative features include affective flattening, poverty of speech, retardation, apathy, lack of sociability. In general, the term relapse is used to refer to a deterioration or re-occurrence of positive rather than negative features. Negative features have been shown to be more stable than positive features (Johnstone *et al.* 1987) and the situation whereby negative symptoms rapidly deteriorate, resulting in the urgent need for readmission to hospital, is rare.

Although it is clear that relapse in schizophrenia does not usually refer to negative features, the boundaries of the term are not otherwise entirely clear. Sometimes, relapses are defined as type I—the reappearance of schizophrenic symptoms in a patient who had been free of them following the initial episode, and type II—the exacerbation of persistent positive symptoms. These types are not always easily differentiated. The experience of being told by a patient in relapse that his positive symptoms have been present at least to some extent for months, even although he has on repeated occasions denied all positive symptoms throughout that period, is not unusual.

It has been considered (Falloon *et al.* 1978) that because of the difficulties of measuring schizophrenic symptomatology it is necessary to supplement such assessments by relatively crude indicators of change, such as admission to hospital. Using this method, these authors found that depression was a frequent cause of readmission to hospital and that this was sometimes but not always associated with positive schizophrenic features. Social difficulty not associated with any change in symptomatology was an uncommon cause of admission in that study.

In this chapter the term relapse is being used to refer to deterioration in positive symptoms which may or may not have been present in a lesser degree before the relapse, but it is recognized that some of the relapses described in the articles cited may represent readmissions due to social difficulty associated with symptoms which have shown no great recent change. The heterogeneity of schizophrenia has received much recent interest (Johnstone in press) and numerous studies have compared and contrasted the course of the disorder in patients with and without a family history of the disorder, a history of obstetric complications or evidence of structural brain changes (Murray *et al.* 1985; Johnstone *et al.* 1989; Lewis *et al.* 1989; Gattaz *et al.* 1990). Although it is suggested that these variables do relate to the course of the illness, this is in the sense of deterioration and the development of cognitive deficits or other evidence of a defect state and not in the sense of risk of relapse. The relationship between schizophrenic relapse and the development of the defect state remains unclear.

Schizophrenic relapse may be influenced by pharmacological and social means.

Pharmacological effects

The value of neuroleptic drugs in the prevention of schizophrenic relapse has been clearly shown. The benefits in this situation of both oral medication (Leff and Wing 1971) and depot parenteral medication (Hirsch *et al.* 1973) are well established, and in a review of 24 controlled studies of maintenance treatment in schizophrenia Davis (1975) concluded that the evidence for such efficacy is overwhelming. Neuroleptics do not, however, prevent relapse. In the study of Hogarty *et al.* (1974), 48 per cent of patients on active neuroleptic medication relapsed within two years and this figure was not substantially different in a later study (Hogarty *et al.* 1979) of depot prophylactic medication (i.e., where compliance was assured). Furthermore, Falloon *et al.* (1978) found oral pimozide to be at least as effective as depot injections of fluphenazine decanoate in preventing relapse in a group of schizophrenic patients discharged from hospital, and that the adminis-tration of pimozide was associated with fewer unwanted effects. Almost all trials have included patients who had suffered more than one psychotic episode and therefore it is possible that patients who are subject to repeated relapse are over-represented. It could be considered that a study of patients following a first psychotic episode would be likely to show a higher propor-tion of patients remaining well without neuroleptics than the 15–20 per cent in the above trials who have remained well on placebo after one to two years. The Northwick Park Study (Crow *et al.* 1986) of 120 cases of first episode schizophrenia showed that 46 per cent of those on active neuro-leptics and 62 per cent of those on placebo relapsed within two years. Somewhat more, therefore, remained well than in studies of the generality of patients, but the number remained disappointing. In this study, the most important determinant of relapse was duration of illness prior to the intro-duction of medication. This result is illustrated in Fig. 1.

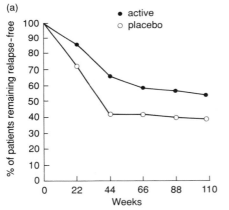

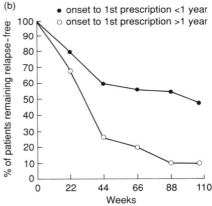

Fig. 1(a). Survival curves of patients in first episodes of schizophrenia study on active and placebo medication.

Fig. 1(b). Survival curves of patients on first episodes of schizophrenia study with pre-treatment intervals of less than 1 year and more than 1 year.

As patients were not randomly allocated to early or late introduction of neuroleptics, it is clearly possible that the two groups differed in some relevant characteristic, but careful scrutiny has not revealed any obvious differences. The finding suggested the tantalizing possibility that the early introduction of medication might have enduring beneficial effects in terms of reducing later relapse rate. This aspect requires further study, although appropriate trials will not be easy to conduct.

Disadvantages of prophylactic neuroleptics

Although the benefits of neuroleptics in reducing the risk of relapse are well established, the long-term use of these drugs is associated with various disadvantages. Both long- and short-term extrapyramidal side-effects may be a problem: the patients frequently dislike the anticholinergic side-effects and the weight gain, and tiredness often associated with maintenance neuroleptic treatment. There is some evidence (Macmillan *et al.* 1986; Johnstone *et al.* 1990) that maintenance neuroleptic treatment may be associated with reduced ability to achieve in occupational and other terms. Compliance is difficult to achieve with any prophylactic medication, but these various problems no doubt contribute to the major difficulties that are frequently encountered in persuading patients to continue with long-term neuroleptics. Several strategies have been adopted to try to minimize these problems. Studies have been conducted with a view to examining the possibility that the dosages currently used are in excess of those required to maintain patients relapse-free.

Low dose neuroleptic prophylaxis

Kane *et al.* (1983) compared low dose (1.25–5 mg fluphenazine fortnightly) with conventional dose (12.5–50 mg fluphenazine fortnightly) depot maintenance treatment and found a relapse rate at one year of 56 per cent in the low dose and 7 per cent in the conventional dose group. Marder *et al.* (1987) compared low dose (5 mg fluphenazine fortnightly) with conventional dose (25 mg fluphenazine fortnightly) maintenance treatment, and found a relapse rate of 36 per cent at two years in the high dose group compared with 69 per cent in the low dose group. Both Kane *et al.* (1983) and Marder *et al.* (1987) found the side-effects to be less in the low dose group. These studies, of course, compared conventional regimes with substantially lower doses and made no attempt to determine the minimum effective dose. This may have been the aim of the study of Branchey *et al.* (1981) where prophylactic neuroleptic medication was reduced stepwise to one-half, one-quarter, and one-eighth of its initial value and then the patients were transferred to placebo and their progress compared with that of a group on unchanged medication. None of the patients relapsed during the reduction phase (although many did once they were on placebo) and therefore no information about the lowest effective prophylactic dose was obtained.

Early detection of relapse

Another question that has been considered is the possibility that the detection of prodromal (and essentially non-psychotic features) of schizophrenic relapse may allow the effective introduction of prophylactic neuroleptics on an intermittent basis (Herz and Melville 1980; Herz *et al.* 1982; Heinrichs and Carpenter 1985). Recent work of Hirsch, Jolley, and their coworkers (Jolley *et al.* 1989, 1990; Hirsch *et al.* 1990) has indicated that the early treatment with neuroleptics of prodromal features of schizophrenic relapse is not an appropriate alternative to continuous prophylactic medication and that maintenance neuroleptic treatment is the better regime.

Alternatives to prophylactic neuroleptics

A recent study conducted by myself and coworkers compared the value as a prophylaxis against relapse of pimozide, lithium, pimozide; and lithium and placebo in a population of patients with functional psychotic illness who had responded to the same treatments during an acute episode of psychotic (and essentially schizophrenic) illness (Johnstone *et al.* 1991*b*). The principal issue in this study, which is relevant to the matter of developing alternatives to prophylactic neuroleptic medication in schizophrenic patients, is the question of the value of prophylactic neuroleptics in patients who had recovered from their acute episode of psychosis without active neuroleptics. It might be thought reasonable to suggest that those patients who can recover from a psychotic episode without dopamine blocking drugs might not require maintenance dopamine blockers in order to remain well, and that therefore recovery from an acute episode without active neuroleptic medication would be an indication that maintenance neuroleptic treatment would not be necessary. It was, however, clear from this invest-

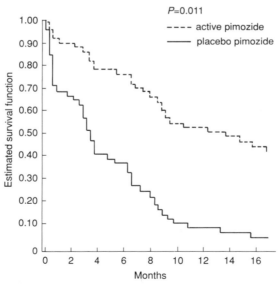

Fig. 2. Survival curves of patients in maintenance phase of functional psychosis trial comparing those whose regime included active and placebo pimozide.

igation that patients who had achieved a set standard of recovery without neuroleptics and who were maintained on a neuroleptic-free regime in the months following recovery did not do well during that time. In spite of the small numbers in the study (30 cases in all) the survival of those who did not receive the neuroleptic pimozide was significantly shorter ($P = 0.01$) than that of those patients who achieved recovery on active neuroleptic treatment and were maintained on the same regime after discharge (Fig. 2).

This study also addressed the issue of the value of lithium as a prophylactic following psychotic and essentially schizophrenic episodes and found that it was of no additional value (Fig. 3).

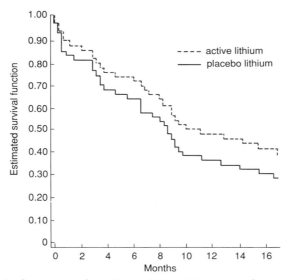

Fig. 3. Survival curves of patients in maintenance phase of functional psychosis trial comparing those whose regime included active and placebo lithium. (No significant difference.)

The use of lithium in acute and chronic schizophrenia has been described in various studies (Delva and Letemendia 1982), but these are concerned with symptomatic treatment and not with prophylaxis. Similar considerations apply to most studies of other non-standard treatments of schizophrenia, such as atypical neuroleptics like clozapine (Kane *et al.* 1988) or drugs of an entirely different kind such as cholecystokinin (Montgomery and Green 1988) or carbamazepine (Johnstone 1987).

Social effects
A number of social factors are known to be associated with exacerbation of positive symptoms in schizophrenic patients. A series of studies conducted

by Wing and coworkers in the late 1950s and early 1960s and reviewed recently by Wing (1989) showed that strong and unrealistic pressure to perform to high occupational and social standards was likely to precipitate relapse. Life changes of various and not necessarily negative kinds are also associated with relapse (Brown and Birley 1968). Later work has concentrated on the concept of 'high expressed emotion' which is most simply operationalized in terms of critical comments made by some key relatives (Wing 1989). This concept has been studied in relation to, and considered relevant for the prognosis of, a wide range of disorders including depression, eating disorders, and inflammatory bowel disease. A number of studies in a variety of cultural settings have demonstrated a significant association between high expressed emotion and schizophrenic relapse (Brown *et al.* 1972; Vaughn and Leff 1976; Leff and Vaughn 1981; Vaughn *et al.* 1984; Moline *et al.* 1985; Karno *et al.* 1987; Leff *et al.* 1987; Tarrier *et al.* 1989), although there are some studies where the association is less clear (Macmillan *et al.* 1986; Dulz and Hand 1986; McCreadie and Robinson 1987; Parker *et al.* 1988).

Studies of educational, supportive and therapeutic groups (Leff *et al.* 1989; Tarrier *et al.* 1989) have usually but not always (Kottgen *et al.* 1984) indicated that it is possible by family intervention of this kind to reduce or at least delay relapse in vulnerable cases (Tarrier *et al.* 1989). Family interventions of this kind are, of course, not associated with the side-effects which may be such a problem with maintenance neuroleptic treatment.

Not all families, however, are willing to co-operate with such management, and indeed those most in need may be least willing to co-operate (Leff *et al.* 1989). Some studies have shown relatively few schizophrenic patients to be living with relatives who exhibit high expressed emotion (EE) (Macmillan *et al.* 1986; McCreadie and Robinson 1987). Furthermore, it is worth bearing in mind that ideas fashionable earlier in the century concerning the possibility that pathological relationships or patterns of communication in the nuclear family lead to the development of schizophrenia cast a long shadow. Although the work of, for example, Lidz *et al.* (1965) and Alanen (1958) did not have experimental support and their ideas have not stood the test of time, concepts of the 'schizophreno-genic mother' and 'double-bind' have been widely discussed in the mass media and continue to cause anxiety to concerned relatives. It is important to allay such concerns when dealing with the concept of expressed emotion.

Conclusions

As far as relapse in schizophrenia is concerned, the most important issue is that the risk of relapse remains high and that few patients in whom this diagnosis is appropriately made will escape it. There is no doubt that maintenance neuroleptics reduce this risk, but they do not eliminate it. The use of this form of management is associated with a number of problems of which the most relevant is probably the fact that very large numbers of patients cannot be persuaded to continue with it. Psychiatrists and other professionals may think that the benefits of the treatment outweigh the disadvantages, but many patients do not agree. The question of the length of time for which a patient should continue on depot medication after a psychotic episode is often raised, but very often this is a decision that the psychiatrist does not have to make because many patients decide for themselves that they wish to stop. Reduction of over-high expectations, social stimulation in general, and a variety of life's experiences may reduce the risk of relapse, but the resulting life-style will be unappealing to less impaired sufferers. Family intervention is likely to help appropriate families who accept this form of management, but there are many patients for whom this is not applicable. Reduction of relapse is an area where even in the best of circumstances disappointments will be commonplace, as all means of intervention currently available are of only limited efficacy. This is a disappointing conclusion, but in my view, while we may help our patients and their relatives by presenting a rather more optimistic picture than a cold appraisal of the facts really allows, it is important that we admit at least to ourselves that this is an area where available methods are not adequate—where therefore we must continue to seek new and better approaches.

Clinical guidelines

It will be clear from the foregoing that it is difficult to provide guidelines for the prevention of relapse in individual patients on the basis of the literature. These guidelines are therefore based upon clinical experience as well as scientific evidence.

1. Most schizophrenic patients will experience relapses. The frequency of these is likely to be reduced by maintenance neuroleptic treatment, but such treatment has unwanted effects and patients do not like taking it. For most patients who have had at least two episodes the aim of continued management is to maintain the patient on a dose of a neuroleptic which has as few side-effects as possible for as long as possible and, if it can be achieved, for at least three years from the last relapse.

2. In order to achieve (1) the nature, dosage, and regime of the neuroleptic and of concomitant anti-Parkinsonian medication should be adjusted to achieve minimal side-effects and the administration of the treatment should be arranged to suit the patient. Some patients prefer to receive injections or tablets from their general practitioner (GP), some are

willing to attend an injection clinic and some like to return to the ward or the staff that they knew during their admission. In my experience, it is well worth respecting the patient's wishes in these matters, and in particular awkward arrangements which, for example, involve collecting a prescription from a GP and ampoules of injection from a pharmacy, and then taking these to a clinic, are best avoided.

3. Although studies demonstrate the value of maintenance neuroleptics following first episode schizophrenic episodes, it is clear that some patients do remain well for years after such episodes, even though they do not receive active medication. It is therefore reasonable to allow patients who have only had one episode a period off medication. Rules for this can only be arbitrary, but the following scheme is practical.

    (a) Patients who have had a relatively brief episode of sudden onset for which treatment was rapidly introduced have a 50 per cent chance of remaining well without neuroleptics. If such patients are reluctant to take maintenance treatment at all, the best form of management may be to agree to this, provided that the patient keeps occasional contact with the service for one year.

    (b) In the generality of first episode cases it is advisable to continue maintenance treatment for one year, but if at the end of that time there have been no problems it is reasonable to withdraw it, provided there are no circumstances which would make this inadvisable.

    (c) In patients who have had a first episode where circumstances are such that a relapse within a particular period of time would have especially adverse consequences (e.g., those who become ill while they are students or are engaging in some other form of study) prophylactic medication should be continued until that time is past.

4. If it is possible and appropriate for family members to participate in a family intervention programme, then this should be arranged. Where this is not possible or acceptable, relatives should be given a clear explanation of the nature of the disorder and the problems that the patient may encounter. They should have ready access to the service and often find it helpful to be given direct advice on the day-to-day management of that individual patient's behaviour. When this is done the likelihood of their seeking admission for the patient in an inappropriate way will be reduced.

5. It is not possible, or indeed desirable, for the psychiatric services to organize all aspects of the patient's life, but advice on issues, such as a suitable choice of work for the patient, is often sought. This, of course, is always limited by practical considerations. The most appropriate form of work is probably one involving a limited degree of skill, the

need for little social interaction or decision-making, and the possibility of holding a number of fairly short-term posts which would not necessarily be of the same level of responsibility. For women, secretarial training may be very suitable. If their illness is benign they may be able to function effectively in skilled and interesting posts. If they do less well but are still able to work, then they may be able to follow the occupation for which they are trained at a much lower level and in a variety of short-term posts through agencies. For men, clerical or semi-skilled work of any kind may be suitable, although it is increasingly difficult to achieve. For schizophrenic patients of both sexes a family business, in which they can be employed at varying levels depending upon their state of health, is a great asset.

6. The aim of the long-term management of schizophrenic patients is to help them to achieve the best level of functioning possible. In order to do this it is usually necessary to persuade the patient to accept that his or her condition may cause continuing or recurrent problems, and indeed relapses, but that these will be reduced by appropriate management. Patients and relatives are often reluctant to accept this in the early stages of the illness and the patient's long-term management will be greatly helped if their co-operation and confidence in the psychiatric services are not undermined at this stage. To achieve this aim it is often worthwhile for the psychiatrist to modify the proposed treatment plan to take account of the wishes, plans, and prejudices of the patient and his or her family.

References

Alanen, Y. O. (1958). The mothers of schizophrenic patients. *Acta Psychiatrica et Neurologica Scandinavica*, **33**, Suppl. 124.

Bleuler, M. (1978). *The schizophrenic disorders* (trans. M. Clemens). Yale University Press, New Haven, CT.

Branchey, M. H., Branchey, L. B., and Richardson, M. A. (1981). Effect of neuroleptic adjustment on clinical condition and tardive dyskinesia in schizophrenic patients. *American Journal of Psychiatry*, **138**, 608–12.

Brown, G. W. and Birley, J. L. T. (1968). Crises and life changes and the onset of schizophrenia. *Journal of Health and Human Behaviour*, **9**, 203–14.

Brown, G. W., Birley, J. L. T., and Wing, J. K. (1972). Influence of family life on the outcome of schizophrenia. *British Journal of Psychiatry*, **121**, 241–58.

Crow, T. J., Macmillan, J. F., Johnson, A. L., and Johnstone, E. C. (1986). The Northwick Park study of first episodes of schizophrenia. II. *British Journal of Psychiatry*, **148**, 120–7.

Davis, J. M. (1975). Overview: maintenance therapy in psychiatry and schizophrenia. *American Journal of Psychiatry*, **132**, 1237–45.

Delva, N. J. and Letemendia, F. J. J. (1982). Lithium treatment in schizophrenia and schizoaffective disorders. *British Journal of Psychiatry*, **141**, 387–400.

Dulz, B. and Hand, I. (1986). Short-term relapse in young schizophrenics. Can it be predicted and affected by family (CFI) patient and treatment variables? In *Treatment of schizophrenia* (ed. M. J. Goldstein, I. Hand, and K. Hahlweg), pp. 59–75. Springer, Berlin.

Falloon, I., Watt, D. C., and Shepherd, M. (1978). A comparative controlled trial of pimozide and fluphenazine decanoate in the continuation therapy of schizophrenia. *Psychological Medicine*, **8**, 59–70.

Fish, F. J. (1962). *Schizophrenia*. Wright, Bristol.

Gattaz, W. F., Kohlmeyer, K., and Gasser, T. (1990). Computer tomographic studies in schizophrenia. In *Search for the causes of schizophrenia*, Vol. II (ed. H. Hafner and W. F. Gattaz), pp. 242–56. Springer, Heidelberg.

Heinrichs, D. W. and Carpenter, W. T. (1985). Prospective study of prodromal symptoms in schizophrenic relapse. *American Journal of Psychiatry*, **142**, 371–3.

Herz, M. I. and Melville, C. (1980). Relapse in schizophrenia. *American Journal of Psychiatry*, **137**, 801–5.

Herz, M. I., Szymanski, H. V., and Simon, J. (1982). Intermittent medication for stable schizophrenic outpatients: an alternative to maintenance medication. *American Journal of Psychiatry*, **139**, 918–22.

Hirsch, S. R., Gaind, R., Rohde, P. D., Stevens, B. C., and Wing, J. K. (1973). Outpatient maintenance of chronic schizophrenic patients with long-acting fluphenazine: double-blind placebo controlled trial. *British Medical Journal*, **1**, 633–7.

Hirsch, S. R., Jolley, A. G., McRink, A., and Wilson, L. (1990). Trial of brief intermittent neuroleptic prophylaxis for selected schizophrenic outpatients: clinical and social outcome at 2 years. *Schizophrenia Research*, **3**, 40–1.

Hogarty, G. E., Goldberg, S. C., Schooler, N. R., and Ulrich, R. F. (1974). Collaborative Study Group. Drugs and sociotherapy in the aftercare of schizophrenic patients. II. Two year relapse rates. *Archives of General Psychiatry*, **31**, 603–8.

Hogarty, G. E., Schooler, N. R., Ulrich, T., Mussare, F., Ferro, P., and Herron, E. (1979). Fluphenazine and social therapy in the aftercare of schizophrenic patients. *Archives of General Psychiatry*, **36**, 1283–94.

Johnstone, E. C. (1987). Physical treatments. *British Medical Bulletin*, **43**, 689–708.

Johnstone, E. C., Crow, T. J., Frith, C. D., Carney, M. W. P., and Price, J. S. (1978). Mechanism of antipsychotic effect in acute schizophrenia. *Lancet*, **i**, 848–51.

Johnstone, E. C., Crow, T. J., Johnson, A. L., and Macmillan, J. E. (1986). The Northwick Park study of first episodes of schizophrenia. 1. Presentation of the illness and problems relating to admission. *British Journal of Psychiatry*, **148**, 115–20.

Johnstone, E. C. (in press). Heterogeneity in Schizophrenia: Clinical and Biological Aspects. In *New biological vistas on schizophrenia* (ed. J. P. Lindenmayer and S. Kay). Brunner Mazel, New York.

Johnstone, E. C., Owens, D. G. C., Frith, C. D., and Crow, T.J. (1987). The relative stability of positive and negative features in chronic schizophrenia. *British Journal of Psychiatry*, **150**, 60–4.

Johnstone, E. C., Owens, D. G. C., Bydder, G. M., Colter, N., Crow, T. J., and Frith, C. D. (1989). The spectrum of structural brain changes in schizophrenia: age of onset as a predictor of cognitive and clinical impairments and their cerebral correlates. *Psychological Medicine*, **19**, 91–103.

Johnstone, E. C., Macmillan, J. F., Frith, C. D., Benn, D. K., and Crow, T. J. (1990). Further investigation of the predictors of outcome following first schizo-phrenic episodes. *British Journal of Psychiatry*, **157**, 182–9.

Johnstone, E. C., Leary, J., Frith, C. D., Owens, D. G. C., and Wilkins, S. (1991a). Assessment of the disabilities and circumstances of patients with chronic schizo-phrenia discharged from Harrow beds between 1975 and 1985. *British Journal of Psychiatry*, **159**, Suppl. 13.

Johnstone, E. C., Crow, T. J., Owens, D. G. C., and Frith, C. D. (1991b). The Northwick Park 'Functional' Psychosis Study. Phase 2. Maintenance Treatment. *Journal of Psychopharmacology*, **5**, 388–95.

Jolley, A. G., Hirsch, S. R., McRink, A., and Marchanda, R. (1989). Trial of brief intermittent neuroleptic prophylaxis for selected schizophrenic outpatients: clinical outcome at one year. *British Medical Journal*, **248**, 985–90.

Jolley, A. G., Hirsch, S. R., Morrison, E., McRink, A., and Wilson, L. (1990). Trial of brief intermittent neuroleptic prophylaxis for selected schizophrenic outpatients: clinical and social outcome at two years. *British Medical Journal*, **301**, 837–42.

Kane, J. M., *et al.* (1983). Low dose neuroleptic treatment of outpatient schizo-phrenics. *Archives of General Psychiatry*, **40**, 893–6.

Kane, J., Honigfeld, G., Singer, J., Meltzer, H. and the Clorazil Collaborative Study Group (1988). Clozapine for the treatment-resistant schizophrenic. *Archives of General Psychiatry*, **45**, 789–96.

Karno, M., *et al.* (1987). Expressed emotion and schizophrenia outcome among Mexican–American families. *Journal of Nervous and Mental Disease*, **175**, 143–51.

Kottgen, C., Sonnichsen, I., Mollenhauer, K., and Jurth, R. (1984). Group therapy with the families of schizophrenic patients: results of the Hamburg Camberwell Family Interview Study, III. *International Journal of Family Psychiatry*, **5**, 84–94.

Kraepelin, E. (1919/1971). *Dementia praecox* (trans. R. M. Barclay, ed. G. M. Robertson). Facsimile edition published by Krieger, New York.

Leff, J. P. and Vaughn, C. (1981). The role of maintenance therapy and relatives' expressed emotion in relapse in schizophrenia: a two year follow up. *British Journal of Psychiatry*, **139**, 102–4.

Leff, J. P. and Wing, J. K. (1971). Trial of maintenance therapy in schizophrenia. *British Medical Journal*, **3**, 599–604.

Leff, J. P., *et al.* (1987). The influence of relatives' expressed emotion on the course of schizophrenia in Chandigarh. *British Journal of Psychiatry*, **151**, 166–73.

Leff, J. P., Berkowitz, R., Shavit, N., Strachan, A., Glass, I., and Vaughn, C. (1989). A trial of family therapy in a relatives' group for schizophrenia. *British Journal of Psychiatry*, **154**, 58–66.

Lewis, S. W. (1989). Congenital risk factors for schizophrenia. *Psychological Medicine*, **19**, 5–13.

Lidz, T., Fleck, S., and Cornelison, A. R. (1965). *Schizophrenia in the family*. International Universities Press, New York.

Marder, S. R., Van Putten, T., Mintz, J., Lebell, M., McKenzie, J., and May, P. R. A. (1987). Low and conventional dose maintenance therapy with Fluphen-azine Decanoate. *Archives of General Psychiatry*, **44**, 518–21.

Macmillan, J. F., Gold, A., Crow, T. J., Johnson, A. L., and Johnstone, E. C. (1986). Northwick Park study of first episodes of schizophrenia IV. Expressed emotion and relapse. *British Journal of Psychiatry*, **148**, 133–44.

McCreadie, R. G. and Robinson, A. D. T. (1987). The Nithsdale schizophrenia survey. VI. Relatives' expressed emotion prevalence, patterns and clinical assessment. *British Journal of Psychiatry*, **150**, 640–4.

Moline, R. A., Singh, S., Morris, A., and Meltzer, H. (1985). Family expressed emotion and relapse in schizophrenia in 24 urban American patients. *American Journal of Psychiatry*, **151**, 314–20.

Montgomery, S. A. and Green, M. C. D. (1988). The use of cholecystokinin in schizophrenia: a review. *Psychological Medicine*, **18**, 593–603.

Murray, R. M., Lewis, S. W., and Reveley, A. M. (1985). Towards an aetiological classification of schizophrenia. *Lancet*, **i**, 1023–6.

Parker, G., Johnstone, P., and Hayward, L. (1988). Parental 'expressed emotion' as a predictor of schizophrenic relapse. *Archives of General Psychiatry*, **45**, 806–13.

Tarrier, N., *et al.* (1989). Community management of schizophrenia. A two year follow up of a behavioural intervention with families. *British Journal of Psychiatry*, **154**, 625–8.

Vaillant, G. E. (1963). The natural history of the remitting schizophrenias. *American Journal of Psychiatry*, **120**, 367–76.

Vaughn, C. E. and Leff, J. P. (1976). The influence of family and social factors on the course of psychiatric illness: a comparison of schizophrenic and depressed neurotic patients. *British Journal of Psychiatry*, **129**, 125–37.

Vaughn, C. E., Snyder, K. S., Jones, S., Freeman, W. B., and Falloon, I. R. H. (1984). Family factors in schizophrenic relapse: a California replication of the British research in expressed emotion. *Archives of General Psychiatry*, **41**, 1169–77.

Wing, J. K. (1989). The concept of negative symptoms. *British Journal of Psychiatry*, **155**(Suppl. 7), 10–14.

14

Affective symptoms and schizophrenia: what are the implications for diagnosis and treatment?

JOHN CUTTING

Introduction

This topic lies at the very heart of the problem of psychiatric diagnosis and treatment, because it undercuts one of the fundamental assumptions of psychiatric treatment from the 1950s onwards: the notion that neuroleptics for schizophrenia, lithium for mania, and antidepressants or ECT for depressive illness are specific treatments. If Kraepelin was correct in there being three functional psychoses, two of which are polar variants of one another, and if neuroleptics, lithium, and antidepressants or ECT *are* effective in curing or largely ameliorating their respective target psychosis, then there is no dilemma at all in the management of this group of psychiatric disorders. The problem merely becomes one of refining the diagnostic criteria to achieve maximum discrimination between the three psychoses, ensuring that these criteria are as 'user-friendly' as possible to take account of the undoubted difficulties in establishing the mental state of a psychotic subject, and then administering the appropriate remedy. A poor response at the end of the process would then be explained by one or more of the following factors: poor compliance by the patient, an unusual biochemical make-up (e.g., fast metabolizer) or some other condition attenuating the efficacy of medication (e.g., drug addiction, alcoholism), a greater than usual diagnostic incompetence on the part of the psychiatrist, or a particularly severe form of the illness.

Unfortunately, the situation is not as clear-cut as this, and there is a dilemma if one starts to question the two basic assumptions: (1) that Kraepelin's diagnostic system is substantially correct; and (2) that each psychosis response to mutually exclusive types of biological treatment. In this chapter these two assumptions will be examined.

Further very practical issues regarding affective symptoms in schizophrenia concern the fact that depression and hopelessness often complicate the course of schizophrenic disorders, and contribute much to the risk of suicide in this condition. It has been estimated that the risk of suicide is nearly as great as that in manic-depressive disorder, with approximately 10 per cent of individuals with schizophrenia taking their own lives (Miles 1977), and the risk being highest during the early years of the illness. However, risk factors for suicide in schizophrenia (p. 180) are somewhat different to those in manic-depressive disorder. The risk of suicide highlights the need for careful surveillance for periods of risk, and instigation of appropriate treatment when they occur.

The manifold nature of affective symptoms and signs

Affective symptoms

If we use the conventional meaning of the term 'symptom' as something that patients complain about to their doctors, and regard affect, emotion, and mood as synonymous, then most 'affective symptoms' are emotional states that a patient does not want to have. They can be so-called 'simple' emotions—elation, misery, anxiety, or more complex—guilt, existential despair. They can even be complaints about having no emotions—anhedonia —so bitterly expressed, sometimes, that it is clear that there *is* a strong emotion about having *no* emotions.

In addition to voiced complaints about unwanted subjective experiences, there may be complaints about another aspect of emotion—drive or motivation. In most definitions of emotion, particularly Ryle's (1949), it is a driving force as well as an experience. Apathy or impaired zest have also, therefore, to be deemed 'affective symptoms'.

There is a tendency among some psychiatrists to regard another set of complaints—altered sleep pattern, disturbed appetite or weight, changes in energy level, impaired concentration—as 'affective symptoms'. I think that this practice is misleading, as the above-mentioned may not be associated with complaints of altered feelings and can occur in all the major psychoses. The alternative term 'biological symptoms' is also misleading, because it does not denote some of the most characteristic symptoms of a depressive illness—reduced concentration, muzzy thinking, impaired memory. The best term, in my view, is impaired vital/executive activity.

Affective signs

If we adhere to a correct definition of the word symptom we are left with another set of phenomena which are sometimes referred to as 'affective symptoms', but which are strictly a set of *signs* that we observe concerning the way patients express their own emotion or perceive the emotional state of others. We may infer that someone is miserable (even though he or she may deny it) because of facial expression or tone of voice. We may judge that the person shows less emotion in voice, gesture, and face than expected in a certain situation—blunting of affect—or that what emotion is shown is inappropriate to the situation—inappropriate affect. We may discover, by means of psychological tests, that the patient finds it hard to judge the emotional state of others: both the tone of voice—angry, surprised, etc.—in a series of tape-recorded extracts of a normal person speaking in a designated tone of voice—aprosody or dysprosody; and others' facial expression —sometimes referred to as prosopaffective agnosia.

The multiplicity of causes of affective symptoms and signs

From the above discussion it is clear that there are many types of affective symptoms and signs. It is also true that each variety has a different constellation of causes. Before dealing with the particular problem of schizophrenia, therefore, we need to briefly survey the chief causes of the main varieties of affective symptoms and signs, without which it is impossible to see the issue of affective illness in schizophrenia in context.

Distinguishing normal from pathological emotions

First, with regard to voiced complaints about unwanted subjective experiences of emotion, it is probably true that most people at some time in their lives experience every emotion that there is, even anhedonia (consider the expressions: 'I was numb with fear' or 'I just felt dead inside when I heard the news'). No emotional experience, therefore, can in itself, be deemed pathological.

What about the duration of an emotional experience? Can this render it pathological? I think not. One can feel guilty for the rest of one's life over an incident in childhood or early adulthood. There are people who, for want of a better term, have always been 'miserable sods' and always will be. There are even cases of otherwise normal individuals with lifelong anhedonia: a study of bomb-disposal personnel in Northern Ireland showed that some of these men had volunteered for the job because they thought it might give them an emotional experience which nothing else in their lives had ever done.

What about the severity of an emotional experience? I doubt if this too can ever render it pathological. The depth of despir, guilt or bereavement is part of the rich pattern of life and certainly not pathological, so unwanted subjective experiences of emotion, whether prolonged or severe, and of whatever variety, can never be specific to any psychopathological condition.

Disorders of drive

The second aspect of disordered emotion, involving drive or motivation, is more relevant to the theme of this chapter, because patients with a depressive illness or schizophrenia, or their relatives, complain of there being reduced drive. The complaint is not pathological in itself, because many normal people experience this under various conditions. It is, however, a prominent feature of both schizophrenia and depressive illness. Although the nature and cause of this phenomenon may be different in different conditions, on a practical level it is a poor discriminator between normal and pathological states of mind, or between the various pathological states themselves. I once saw a patient who had been lying in bed for five years. When I asked him why, he replied that he had been studying palmistry, and proceeded to give a masterly demonstration of the art. There was nothing else to give me a clue as to whether he was a mere eccentric, depressed or schizophrenic.

'Biological' affective symptoms

What about the so-called 'biological symptoms', viewed as affective symptoms? Here, I think there is a diagnostic usefulness. But the critical feature, in this respect, is not the presence or absence of a package of these, but the particular quality of each. Moreover, the diagnostic specificity of these symptoms is in the distinction between normality and an affective psychosis, not, in my view, between an affective psychosis and schizophrenia. For example, in deciding if someone is understandably miserable as a result of a life event, or suffering from a physiologically determined depressive illness, I place a lot of weight on whether there is a triad of reduced zest for life, lowered energy and impaired concentration, coupled with a worsening of all these in the morning. In deciding if someone is hypomanic or merely an exuberant personality, I regard a speeding up of speech, thinking and activity, coupled with markedly reduced sleep, as the most helpful variables. But either slowing down or speeding up of mental activity is very common in schizophrenia, particularly in the early states of the condition, and are of little diagnostic use in themselves. This alteration in vital or, as I prefer to call it, executive activity of the mind is characteristic of an affective psychosis, but is not pathognomonic. The schizophrenic state of mind may or may not contain a *quantitative* change in the level of activity, but what characterizes it is the *additional* presence of an entirely different set of phenomena which stem from a *qualitative* change in the way the mind works.

Affective symptoms and the diagnosis of schizophrenia

In my view, affective symptoms, or even affective signs, such as blunting of or inappropriate affect, are the least reliable way of identifying the crucial qualitative difference between affective illness and schizophrenia. Take blunting of affect. This is a composite judgement on the part of an observer, based on tone of voice, gesture, facial expression, and level of muscular activity. But a depressive illness can affect all these variables, as can neuroleptics, as well as schizophrenia. Take inappropriate affect. Manic patients, according to Kraepelin (1921), have 'movements of expression which are for the most part very vivacious. The patient makes faces, rolls his eyes, assumes theatrical attitudes, stands erect, salutes in military fashion'. These are generally inappropriate to the situation.

In fact, of all the affective symptoms and signs listed earlier, the most diagnostically specific are those which involve the ability of a psychotic subject to appreciate the emotional state of someone else: schizophrenics are significantly inferior to those with a depressive illness in identifying others' facial expressions (Gessler *et al.* 1989) and inferior to both manics and those with a depressive illness in recognizing the emotional content of others' speech (Murphy and Cutting 1990). Unfortunately, these aspects of emotion are difficult to test in an acutely psychotic subject, and the variance is large.

Diagnostic categories

From the foregoing comments it should be clear that I regard affective symptoms not only as misleading in distinguishing between mania or depressive psychosis and schizophrenia, but as a red herring. On the theoretical level, diagnostic reliance on their presence has led to such artificial concepts as schizoaffective psychosis, unitary psychosis, and various continental terms such as 'affect-laden paraphrenia' or 'cycloid psychosis' as described below.

Schizoaffective psychosis

Brockington's two studies (Brockington *et al.* 1980*a*,*b*) contributed to the decline in use of the term schizoaffective psychosis. In these studies, he and his colleagues found that schizomanic patients were virtually indistinguishable from typical manics (in background features, course, and response to treatment) and that schizodepressives were a heterogeneous collection of psychotic patients. Neither study confirmed the nosological independence of a schizoaffective psychosis, and genetic studies from the US (Cohen *et al.* 1972; Pope *et al.* 1980) found it to be genetically heterogeneous as well.

Unitary psychosis

The vogue for diagnosing a unitary psychosis has waxed and waned since Griesinger (1845) first proposed it. In the UK it is currently in fashion again, partly due to Crow's (1986) espousal of the concept, and partly through the tendency of some clinical psychologists (Eysenck 1972; Claridge 1972) to view psychosis as a dimensional attribute of personality or as a hierarchical process of mental breakdown (Foulds 1965). However, I do not find the notion of a unitary psychosis compelling. To think that the mind has only one way, with minor variations, of going mad regardless of cause or site of brain damage seems equivalent to the rather naïve views of those early twentieth-century neurologists (e.g., von Monakow and Mourgue 1928; Lashley 1929) who disputed Broca's ideas on localization of aphasia.

Paranoid psychosis

Another false trail to which the ubiquitous presence of affective symptoms in psychosis has led is the plethora of terms for a paranoid psychosis with marked affective features. Affect-laden paraphrenia (Leonhard 1957/1979), ego psychosis with autochthonous ideas (Wernicke 1900), and *Angstpsychose* (anxiety psychosis, Wernicke 1900) are examples. It is certainly true that, practically, it is very difficult to distinguish a paranoid state devoid of any schizophrenic features but with accompanying misery or elation (delusional disorder in the DSM-III-R nomenclature) from an affective psychosis with marked paranoid delusions. Erotomania is a good example. I once saw a patient, a man of 40, who had held the belief for the previous three months that his parents were conspiring to marry him off to a tennis star. The referring psychiatrist had regarded this as a paranoid psychosis as there were additional delusions of reference. When I interviewed him, however, he admitted to racing thoughts, insomnia, and over-activity at the height of his belief, along with the belief that the tennis star had specifically looked at him among the crowd after winning a match in one of the tournaments leading up to Wimbledon. I considered that his illness was manic, because of evidence of the speeding up of mental processes.

Cycloid psychosis

What about cycloid psychosis, a term coined by Kleist (1928/1974) and refined by Leonhard (1957) and Perris (1974)? I used to think (Cutting *et al.* 1978) that of all the terms proposed for a psychosis dominated by an admixture of so-called schizophrenic phenomena and affective symptoms this had the best claim on our attention. I believed this because its diagnostic criteria, in Perris' (1974) scheme, seemed to distil the best features from all the other atypical psychoses proposed during the century (upwards of 40) and to maximize the difference from both a typical affective psychosis and schizophrenia.

Added to this, cycloid psychosis appeared to pursue a course which was different from any of the typical major psychoses, and, in my experience, not to respond to the specific biological treatments for any of the typical psychoses. Relevant to the present theme, the affective symptoms characteristically fluctuated, unlike the supposed pattern of the elation in mania or the misery in depressive illness which are purportedly constant throughout a single episode.

I now believe that I was wrong to regard cycloid psychosis as what some people have called a typical 'atypical psychosis' or the 'third psychosis'.

I think I was wrong because I was taken in by the claims made for the significance and specificity of the affective symptoms in cycloid psychosis. Its proposed diagnostic features, besides fluctuating mood, comprise perplexity, ecstasy, anxiety, and a disturbance in the speed of execution of movement. All these may occur in an affective psychosis, particularly mania, which is well known for the fluctuation in mood shown by the patient. I therefore now believe that cycloid psychosis, as Kleist originally claimed, is an affective psychosis, where mood fluctuates and changes.

In summary, the diagnostic problem of admixtures of 'schizophrenic' and 'affective symptoms' will continue to plague us until a properly validated diagnostic scheme is developed. This will inevitably require psychological *and* neuropsychological criteria. In the meantime, my advice is to avoid the use of 'schizo-affective' and other 'atypical' labels, and concentrate on abnormalities in vital energy (enhanced or reduced) in diagnosing mania and depression, and look for evidence of disturbed experiences of self, body, mind, and outside world in diagnosing schizophrenia.

Relevance of affective symptoms in drug treatment

Although the efficacy of what I might call the primary treatment options for each of three psychoses was established by the 1960s—neuroleptics for schizophrenia (Cole *et al.* 1964), lithium for mania (Schou 1963), and antidepressants or ECT for depressive illness (MRC 1965)—their specificity in this respect, relative to their efficacy in either of the other two psychoses, has since been questioned. In fact, one can find evidence for the efficacy of every non-primary treatment option, with the exception of antidepressants in schizophrenia.

A further problem arises if one examines the efficacy of the primary treatment option in instances of the target psychosis where there is an admixture of features of one of the other two psychoses. Brockington *et al.* (1978) found, for example, that what they called schizodepressives responded better to chlorpromazine than to amitriptyline, and Glassman *et al.* (1975) found that delusional depressives responded less well to imipramine than did non-delusional depressives. Schizomanics in the study of Brockington *et al.* (1978) study responded as well to lithium as they did to chlorpromazine.

Lack of treatment specificity

Claims as to the specificity of the primary treatment option in each of the three psychoses look distinctly shaky in view of all this (e.g., Schou 1963, for lithium in mania and imipramine in depressive illness as 'mood-normalizers'; supporters of the dopaminergic hypothesis in schizophrenia who use the specificity of neuroleptics to back up this view). A recent study by Johnstone *et al.* (1988) adds further doubt: 120 psychotic individuals, divided into those with elevated mood, depressed mood or no predominant mood disturbance, were given pimozide, lithium, pimozide and lithium or placebo under double-blind conditions over four weeks. The investigators concluded that pimozide significantly improved positive symptoms 'whatever the mood state of the patient' but had no effect on elevated mood, depressed mood or negative symptoms; and that lithium significantly reduced elevated mood but had no effect on depressed mood, positive symptoms or negative symptoms. This is an extremely impressive study, and undermines both neuroleptics' status as treatment-specific for schizophrenia and lithium's status as treatment-specific for mania, although the investigators admit that their study undermines the latter's status less than it does the former's.

Role of specific antidepressant treatment

Depressive symptoms are common in schizophrenia and many patients receive treatment with antidepressant drugs at some stage of their illness. In general, the evidence from controlled trials suggests that adding tricyclic antidepressants to the neuroleptic medication of patients who have active psychotic symptoms is not helpful and may even retard resolution of positive symptoms (Kramer *et al.* 1989). In contrast, Siris *et al.* (1987), in a placebo-controlled trial, found that imipramine combined with maintenance neuroleptic treatment was of significant benefit in patients where depressive episodes occurred after the resolution of active psychotic symptoms. This finding may be of some importance in view of the risk of suicide in such patients (see below).

Prevention of suicide

As already noted, people with schizophrenia are at considerable risk of suicide. This risk is associated with a tendency to develop affective symptoms. So who is most at risk and what can be done to prevent suicide?

On the basis of a comparison of 30 chronic schizophrenics who had killed themselves, with 30 surviving schizophrenics, Roy (1982) identified the following distinguishing characteristics of the suicides: male sex; relatively young age; chronic illness marked by relapses and remissions; depression noted at the last contact; expression of suicidal ideas; and unemployment and recent discharge from in-patient care. However, clearly, these characteristics apply to a large number of schizophrenic patients. Further workers have identified other and more subtle risk factors (Drake *et al.* 1984). One important finding was that all of a series of schizophrenic suicides occurred during a relatively non-psychotic phase of the illness. Furthermore, most of the suicides had attained a relatively high level of educational status prior to the illness, were usually living alone, and had previously indicated suicidal intent. They tended to have high, non-delusional expectations of themselves, were largely aware of the effects of their illness and its consequences for their future functioning, and tended to be depressed, particularly with feelings of inadequacy, hopelessness, and suicidal ideas, rather than biological features of depression. In keeping with the known strong association between suicidal risk and negative expectations about the future, three variables were especially important in predicting suicide, namely: fears of mental disintegration; suicide threats; and hopelessness. Virtually all of the suicides and very few surviving controls had shown some or all of these symptoms.

These findings emphasize the need for patients with schizophrenia to receive long-term support and for careful surveillance to be maintained during remission as well as during relapse, in order to identify depressive symptoms and thinking should these arise. Management of depressive episodes will depend on:

(1) ensuring that the symptoms are not the prodrome of a further psychotic episode;
(2) establishing that any prophylactic neuroleptic medication is being complied with and that the dose is neither too great nor too small;
(3) provision of psychological support, including specific problem-solving measures in order to help the patient deal with difficulties that he or she may be facing as a result of the illness; and
(4) investigation of appropriate antidepressant treatment should this be warranted (see above).

Clinical guidelines

1. The coexistence of affective and psychotic symptoms has important implications for diagnosis and treatment. The starting point of management is a thorough clinical assessment paying particular attention to alterations in vital energy in the detection of altered mood. Remember that in schizophrenia, all kinds of affective disturbance can be present but that a schizophrenic illness is typified by additional qualitative abnormalities in experiences of the mind and body and in their relation with the external world.

2. At present, the basis of drug treatment in schizophrenia is with neuroleptic drugs; in many patients affective symptoms will remit as the psychosis improves. In patients with persistently elevated mood, lithium treatment, co-administered with neuroleptics, may be beneficial.

3. Depressive symptoms need to be distinguished from blunting of affect and the extrapyramidal side-effects of neuroleptics. Specific anti-

depressant drugs seem to be of little benefit in patients with active psychotic symptoms but may be very helpful in combination with neuroleptics in patients where a depressive disorder presents following the resolution of psychosis.

4. The assessment of depression in patients with schizophrenia is particularly important in view of the high risk of suicide (10 per cent) in this group. Young men with a chronic illness and a high educational attainment prior to illness are at particular risk, especially in the period of remission that follows an acute psychotic relapse. Such patients need long-term follow-up with a programme of treatment that focuses on their social and psychological needs, as well as the appropriate use of medication.

References

Brockington, I. F., Kendell, R. E., Kellett, J. M., Curry, S. H., and Wainwright, S. (1978). Trials of lithium, chlorpromazine and amitriptyline in schizoaffective patients. *British Journal of Psychiatry*, **133**, 162–8.

Brockington, I. F., Wainwright, S., and Kendell, R. E. (1980*a*). Manic patients with schizophrenic or paranoid symptoms. *Psychological Medicine*, **10**, 73–83.

Brockington, I. F., Kendell, R. E., and Wainwright, S. (1980*b*). Depressed patients with schizophrenic or paranoid symptoms. *Psychological Medicine*, **10**, 665–75.

Claridge, G. (1972). The schizophrenias as nervous types. *British Journal of Psychiatry*, **121**, 1–17.

Cohen, S. M., Allen, M. G., Pollin, W., and Hrubec, Z. (1972). Relationship of schizoaffective psychosis to manic depressive psychosis and schizophrenia. *Archives of General Psychiatry*, **26**, 539–46.

Cole, J. O., Klerman, C. L., and Goldberg, S. C. (1964). Phenothiazine treatment of acute schizophrenia. *Archives of General Psychiatry*, **10**, 246–61.

Crow, T. J. (1986). The continuum of psychosis and its implication for the structure of the gene. *British Journal of Psychiatry*, **149**, 419–29.

Cutting, J., Clare, A. W., and Mann, A. H. (1978). Cycloid psychosis: an investigation of the diagnostic concept. *Psychological Medicine*, **8**, 637–48.

Drake, R. E., Gates, C., Cotton, P. G., and Whitaker, A. (1984). Suicide among schizophrenics: who is at risk? *Journal of Nervous and Mental Disease*, **172**, 613–17.

Eysenck, H. J. (1972). An experimental and genetic model of schizophrenia. In *Genetic factors in schizophrenia* (ed. A. R. Kaplan), pp. 504–15. Charles C. Thomas, Springfield.

Foulds, G. A. (1965). *Personality and personal illness*. Tavistock, London.

Gessler, S., Cutting, J., Frith, C. D., and Weinman, J. (1989). Schizophrenic inability to judge facial emotion: a controlled study. *British Journal of Clinical Psychology*, **28**, 19–29.

Glassman, A. H., Kantor, S. J., and Shostak, M. (1975). Depression, delusion and drug response. *American Journal of Psychiatry*, **132**, 716–19.

Griesinger, W. (1845). *Mental pathology and therapeutics* (trans. 1867). New Sydenham Society, London.

Johnstone, E. C., Crow, T. J., Frith, C. D., and Owens, D. G. C. (1988). The Northwick Park 'functional' psychosis study: diagnosis and treatment response. *Lancet*, **ii**, 119–25.

Kleist, K. (1928/1974). Über zykloide, paranoide und epileptoide Psychosen und über die Frage der Degenerationspsychosen (trans. 1974). In *Themes and variations in European psychiatry* (ed. S. R. Hirsch and M. Shepherd), pp. 297–331. Wright, Bristol.

Kraepelin, E. (1921). *Manic-depressive insanity and paranoia.* Churchill Livingstone, Edinburgh.

Kramer, M. S., *et al.* (1989). Antidepressants in 'depressed' schizophrenic inpatients. *Archives of General Psychiatry*, **46**, 922–8.

Lashley, K. S. (1929). *Brain mechanisms in intelligence.* University of Chicago Press.

Leonhard, K. (1957/1979). *The classification of endogenous psychoses* (trans. 1979). Irvington, New York.

MRC (Medical Research Council) (1965) Clinical trials of depressive illness. *British Medical Journal*, **i**, 881–6.

Miles, C. P. (1977). Conditions predisposing to suicide: a review. *Journal of Nervous and Mental Disease*, **164**, 231–46.

Murphy, D. and Cutting, J. (1990). Prosodic comprehension and expression in schizophrenia. *Journal of Neurology, Neurosurgery and Psychiatry*, **53**, 727–30.

Perris, C. (1974). A study of cycloid psychoses. *Acta Psychiatrica Scandinavica*, Suppl. 253.

Pope, H. G., Lipinski, J. F., Cohen, B. M., and Exelrod, D. T. (1980). 'Schizoaffective disorder': an invalid diagnosis? *American Journal of Psychiatry*, **137**, 921–7.

Roy, A. (1982). Suicide in chronic schizophrenia. *British Journal of Psychiatry*, **141**, 171–7.

Ryle, G. (1949). *The concept of mind.* Penguin, Harmondsworth, Middlesex.

Schou, M. (1963). Normothymotics, 'mood-normalizers'. *British Journal of Psychiatry*, **109**, 803–9.

Siris, S. G., Morgan, V., Fagerstrom, R., Rifkin, A., and Cooper, T. B. (1987). Adjunctive imipramine in the treatment of postpsychotic depression. *Archives of General Psychiatry*, **44**, 533–9.

von Monakow, C. and Mourgue, R. (1928). *Introduction biologique a l'étude de la neurologie et de la psychopathologie.* Alcan, Paris.

Wernicke, C. (1900). *Grundriss der Psychiatrie.* Thieme, Leipzig.

15

Mentally ill mothers and their babies: what are the benefits and risks of joint hospital admission?

R. CHANNI KUMAR

Introduction

The UK has played a leading role in both research into and treatment of puerperal mental illnesses. The tradition of joint admission of mother and baby to psychiatric units is largely unknown outside the UK, Australia, Canada, and New Zealand (Lindsay and Pollard 1978; Margison and Brockington 1982; Stewart 1989; Buist *et al.* 1990). The rationale behind joint admission can be traced back to the influential work of Bowlby and others on attachment. The aim is to try to preserve and to facilitate the relationship between mother and child, often despite the presence of very severe maternal mental and behavioural disorganization. Babies are usually admitted with their mothers into general adult psychiatric wards in which one or more rooms may flexibly be used as nurseries or bedrooms. Such services have, typically, been carved out of existing facilities without any special additional resources or staff (see Kumar *et al.* 1986). There exist, in addition, a few large, specialized mother and baby units with facilities and part- or full-time staff entirely devoted to the care of severely mentally ill mothers together with their babies (see Margison and Brockington 1982; Kumar *et al.* 1986; Aston and Thomas 1987; Prettyman and Friedman 1991). Such units serve a very important role in taking on difficult second-ary or tertiary referrals, in carrying out research and in education. They are akin to intensive care units, requiring high staff–patient ratios, and are therefore expensive to maintain. For this reason they are vulnerable when health authorities seek to make economies because it is, of course, possible to nurse mentally ill women apart from their infants. There do not seem to have been any large-scale surveys reported of facilities for mothers and babies outside the UK and systematic comparisons of such services in different social and cultural settings are therefore not yet possible.

Most mothers make excellent recoveries from their illnesses and they are almost always glad not to have been separated from their babies. However, no study has been carried out to compare the consumers' perceptions of the management of their illnesses in the two systems, one based upon temporary separation of the mother from her infant and the other upon joint admission. Similarly, there have been no systematic comparisons of 'outcome' in which a number of indices have been examined, including measures of recovery in the mother, the quality of the mother–child relationship, the relationship between the father and the child as well as that between the mother and the father, and finally, measures of the psychological development of the child.

Why has the practice of joint admission not spread from the UK and parts of the Commonwealth, at least to other developed countries? Margison and Brockington (1982) in their excellent review noted that babies were admitted with their mothers in a few places in the US (Schuurmans 1966; Van der Walde *et al*. 1968; Luepker 1972; Grunebaum *et al*. 1975), France (Racamier *et al*. 1961), Switzerland (Masson *et al*. 1977), and Israel (Mester *et al*. 1975). There do not seem to have been any reports of new developments and some of the services, for example in the US, are known to have been discontinued. What has led the Americans and probably the French and others to abandon their experiments with joint admission? Is it just cost? The development of mother and baby units (see Baker *et al*. 1961; Glaser 1962; Fowler and Brandon 1965; Bardon *et al*. 1968; Margison and Brockington 1982) more or less coincided with the expansion and consolidation of psychiatric services in the UK National Health Service (NHS). Was this fortuitous and had there not been an NHS would market forces have stifled the 'baby' at birth? Or are there important debits in the clinical balance sheet, such as possible harm or injury to the infant, or long-term adverse effects of early institutional care? Given the lack of reliable data upon which policies on the care of mothers and babies might have been based, were there important differences in attitudes to motherhood, to mental illness, and to both together, which may have influenced the clinical approach to puerperal disorders in the UK as opposed to other Western nations? Legislation on infanticide certainly differs between nations, with the UK adopting an explicitly humanitarian approach; a survey by the author (unpublished) suggests that in most European countries mentally ill mothers who kill their young children are treated leniently. English law singles out infanticide as an offence which may be treated as a special case of diminished responsibility. In the US, the position varies from state to state and it is not uncommon to find mothers serving punitive prison sentences while their counterparts in the UK are typically placed on probation. Without a more detailed study of clinical practice and of the medico-legal situation regarding puerperal illness and infanticide in different countries, it is not possible to go beyond a juxtaposition of anecdotal observations. Translating anecdotes into hard data will, however, be very difficult. For example, statistics about a comparatively striking offence such as infanticide are probable underestimates, the ways in which suspected cases, indictments, and verdicts are recorded may differ from one legal system to another and the plea, if any, of diminished responsibility may also be handled differently. Documenting similarities and differences between the organized responses of different societies to infanticide would be of interest in itself. In the same vein, any attempt to record the variety of clinical responses to mentally ill mothers and their babies must take into account not only the medical and other facilities that are available, but also the social customs surrounding normal and uncomplicated childbirth.

What do we really know about clinical practice in England and Wales? Two recent surveys of facilities (Kumar *et al.* 1986; Aston and Thomas 1987) have been mentioned, but how do the services actually function with regard to mother and baby? Can one identify hospitals which routinely admit severely mentally ill mothers but *not* their infants. Separation is not 'all or none' and, anyway, the context in which a mother is nursed (separately or together with her baby) may exert a greater influence on outcome than the fact by itself of separation or nursing together. The motivation, attitudes, and specialized knowledge of staff may make a considerable positive contribution to outcome. But consider an opposite argument; well-meaning and highly motivated staff in an institution may actually hinder the child's psychological development and normal tendency to form a progressively stronger attachment to its mother. Thus, in any comparisons of outcome it will be necessary first to define accurately the milieu into which a mother is admitted with or without her baby, to note the amount and nature of contact they have, as well as to note the contribution of the alternative caretakers, be they staff, relatives or foster-parents. In addition, it will be necessary to select sensitive and reliable objective measures of outcome, but it would be unwise to ignore subjective reactions of the consumers, such as satisfaction with the clinical service. It will also be necessary to ensure that like is being compared with like in terms of maternal, clinical, and other characteristics. Thus, in order to prepare the ground for this process of comparison it is necessary briefly to review the size and nature of the clinical problem.

The clinical problem

There are three main conditions which are associated with childbirth: *the maternity blues*, *postnatal depression*, and *post-partum psychosis*. It is surprising how often the names are misapplied, for example, a floridly psychotic mother may be described as 'a bad case of the blues' or someone who is depressed may be referred to as a case of post-partum psychosis. To a considerable extent, such confusion has a real basis because the boundaries between these conditions have not been clearly demarcated and another problem is that there is no obvious rationale behind the way the individual adjectives referring to childbearing have been linked with a particular psychological state; all three conditions could just as easily be qualified by any one of the descriptive terms, indeed a fourth term, 'puerperal', is often used interchangeably with 'post-partum' in relation to psychosis. The puerperium is generally regarded as being of six weeks duration but psychiatrists have borrowed the term from obstetrics and then used it to describe an association between events without, until recently, too much thought about its temporal limits (see discussions by Brockington *et al.* 1982; Kendell *et al.* 1987).

Maternity blues

The maternity blues are transient emotional disturbances, typically occurring on or around the fourth or fifth day post-partum and usually lasting for a few hours or, at most, a day or two. By definition, the blues are dysphoric reactions, although they may be intermingled with normal feelings of happiness and achievement which follow the birth. The blues are very common, occurring in up to 75 per cent of mothers. This high prevalence, when taken together with their mildness and brevity, means that the blues are 'normal' both in the statistical and clinical sense, that is, they are not illnesses or disorders (see the review by Stein 1982). The problem lies in picking out prodromal symptoms and signs of mental illness from the ubiquitous and self-limiting mild mood changes of the blues. It is surprising how often early warning signs, such as perplexity, remarks that seem odd or out of character, unexpected eccentricities of behaviour, are dismissed as being part of the normal blues and it is only when the condition becomes florid that attempts begin to be made to try to seek psychiatric consultation and help.

Postnatal depression

Although the problems posed by postnatal depressive neurosis are obviously less acute and immediately less demanding than those of psychotic breakdowns, they merit at least as much attention. Postnatal depression is a hundred times more common than post-partum psychosis; several surveys have shown that 10 to 15 per cent of women are clinically depressed in the first three months after delivery (see reviews by Kumar 1982; O'Hara and Zekoski 1988) and of them, only two or three out of every hundred are referred to psychiatrists (Nott 1982). Two recent surveys (Meltzer and Kumar 1985; Kendell *et al.* 1987) have shown that about 50 per cent of the women who were admitted up to a year after the birth (three months in the Edinburgh study by Kendell *et al.* 1987) were suffering with non-psychotic psychiatric disturbance—depressive disorders, anxiety states, etc. The admissions may have been necessary in some cases because the depressions were very severe, but it should be noted that many of the mothers were recorded as suffering from minor depression (research diagnostic criteria—RDC, Spitzer *et al*. 1978). Some may have been in the midst of social crises, while others may have been single and isolated or experiencing severe difficulties in relating to their babies. There do not seem to have been any surveys reported in the UK in which depressed mothers were admitted without their babies and it may be necessary, therefore, to draw comparisons with such samples of women taken from other countries, such as Scandinavia and the US.

There is a general consensus that the clinical manifestations of postnatal depression are the same as depression which occurs in other settings, except that the mother's ruminations of guilt and inadequacy feed her worries about being an incompetent and inadequate parent. Her tiredness and exhaustion may exacerbate the real difficulties she has in maintaining a sensitive and expanding dialogue with her baby and her low spirits and low tolerance to stress and frustration may actually make her irritable and hostile to others, including the baby. Very often, such feelings are dealt with by withdrawing from the source of stress. For the baby, the mother's retreat inevitably results in a lack of stimulation and contact. Enforced separation of a depressed mother from her infant is likely to induce even more guilt and feelings of inadequacy. Many such women refuse the offer of in-patient treatment because they do not want to be apart from their babies. Thus, simple comparisons of superficially similar populations of hospitalized depressed mothers, some who were separated from and others who remained with their babies, may be a little misleading.

In concluding this section, two facts must be borne in mind: first, that women with non-psychotic affective disorders account for up to half the admissions in the first postnatal year, usually occurring later than the admissions for psychosis (Meltzer and Kumar 1985); and second, that they are, however, only a very small proportion of the mothers who are depressed after childbirth. Thus, in addressing the question whether or not joint admission of mother and baby is or is not appropriate it is important that indications for admission for non-psychotic illnesses are considered separately from the post-partum psychoses and then the question is addressed about the special needs of the non-psychotic mothers and whether they are best met in the community or by admission with or without the baby.

Post-partum psychosis

Incidence   Childbirth marks a time in their lives when women are most at risk of becoming severely mentally ill. The incidence of post-partum psychosis has usually been evaluated from studies of female admissions to psychiatric hospitals in the weeks and months following delivery and the rate is typically found to be between one and two per 1000 live births. Recent studies indicate that the great majority of cases have an onset of illness within a few weeks or, at most, three months post-partum (Brockington *et al.* 1982). Estimates of the relative risk of psychiatric breakdown (admission) show that in the three months following delivery, a woman is 16 to 20 times more likely to require admission than in an equivalent time prior to conception (Paffenberger 1964; Kendell *et al.* 1981). Primiparae are most at risk and Kendell *et al.* (1987) estimated the relative risk of psychotic breakdown for them in the first month post-partum to be 35. This is about 10 times higher than estimated risks of psychiatric illness associated with other major life events such as bereavement.

Nature of the illness   There is a continuing debate about whether post-partum psychotic illnesses have a distinctive phenomenology which, if demonstrated, would support the case for regarding these conditions as a clinical entity, linked to childbirth as the major aetiological agent (see Brockington *et al.* 1982; Hamilton 1982; Meltzer and Kumar 1985; Kendell 1985; Inwood 1985; Kendell *et al.* 1987). Despite the lack of evidence one way or another 'puerperal psychosis' has been removed from official diagnostic and classification systems (DSM-III-R and ICD 9) and by hindering case-identification this has had serious consequences for research (Meltzer and Kumar 1985). Recent studies in which standardized and operational criteria for diagnosis have been used, as well as restricted temporal criteria for the putative link with birth, have all shown that the post-partum psychoses are affective or schizo-affective in nature (Brockington *et al.* 1981; Dean and Kendell 1981; Meltzer and Kumar 1985; Kendell *et al.* 1987). Only a small minority of mothers are diagnosed as schizophrenic and some of them may have had pre-existing illnesses.

   A condition which does not have a name is not likely to attract resources, and claims against health insurance which include extra payments for special services and facilities for mothers and babies are not likely to be enthusistically greeted by the insurers. Private health care systems based on insurance schemes (e.g., in the US or in continental Europe) have obviously not regarded mother and baby units as potentially profitable despite the fact that the risks and payments can be quite precisely calculated and costed. Several investigators have commented that rates of admission to mental hospital, expressed as a proportion of total births, consistently show that between one and two per 1000 of parturient women are affected (e.g., Hemphill 1952; Polonio and Figueiredo 1955; Murphy 1982; Stern and Kruckman 1983; Kumar 1990). Even allowing for obvious sources of error, it is remarkable that this rate has remained broadly the same for over a century.

Post-partum psychoses often have a stormy and fluctuating course but one of the most rewarding features of these illnesses is the completeness of recovery in the majority of women. Clinical management typically involves hospital admission, and in the UK this is very often with the baby. The treatments are conventional pharmacotherapy and/or ECT. There have been no controlled trials of hormone therapy (e.g., of sex steroids) or of ECT as a treatment of choice. The duration of the illness as judged by the in-patient stay varies with the nature of the symptoms. In a survey by Meltzer and Kumar (1985) it was found that women with manic disorders remained in hospital about twice as long as those who were depressed, on average two months and one month respectively, and the severity of depression did not seem to influence the length of stay in the latter group. Schizophrenic and schizoaffective illnesses were associated with slightly longer stays than for manic disorders. The findings of the study by Meltzer and Kumar (1985) are of interest because all the mothers were admitted with their babies but into a range of facilities in psychiatric hospitals in one health region so that bias due to practice in one particular hospital is likely to have been evened out. Unfortunately, neither Dean and Kendell (1981) nor Kendell *et al.* (1987) reported on duration of hospital in-patient stay in their samples of Scottish mothers. In Edinburgh, mothers are admitted *with* their infants into general psychiatric wards because there are no special facilities for babies. Given the close similarities between the Edinburgh and South East England samples it may be well worth mounting systematic comparisons of the two groups of patients, who probably differ relatively little in terms of social, demographic, and clinical psychiatric factors, or in terms of the therapies provided. However, in Edinburgh most mothers are managed in general adult wards with *ad hoc* arrangements for the infants as in some hospitals in London and the South East (Kumar *et al.* 1986), and in other places they are admitted into specialized units, such as at the Bethlem Royal Hospital, which place an emphasis on maintaining them together with their babies. Except in places with both obstetric and psychiatric case registers, it is no longer possible to determine how many and what kinds of patients are admitted separately from their babies and this important third group is vital for the proposed comparison. Confusion over the nosological status of post-partum mental illness is reflected in the coding guidelines that are in the glossary of ICD 9, with the result that the clinical returns relating to post-partum illness are worse than useless (Meltzer and Kumar 1985). It is not clear how best to identify samples of women managed separately from their infants but otherwise treated in broadly similar ways.

Pre-existing psychiatric disorders and assessment of 'safe' parenting

About a quarter of psychiatric admissions in the puerperium are in cases where the mothers have a definite history of previous mental illness (Meltzer and Kumar 1985; Kendell *et al.* 1987). In their Edinburgh study Kendell *et al.* (1987) found that women with prior histories of manic-depressive psychosis (manic or circular) were at greatest risk of admission in the three months following childbirth (21.4 per cent), whereas those with a history of depressive psychosis had a 13.3 per cent rate of admission after childbirth. The rate of admission for women with schizophrenia was 3.4 per cent, for women with prior histories of depressive neurosis it was 1.9 per cent, and for all other diagnoses the rate of admission was 4.1 per cent.

Typically, manic-depressive patients are readmitted because they suffer an acute puerperal relapse (Dean and Kendell 1981; Brockington *et al.* 1981; Meltzer and Kumar 1985). The survey by Kendell *et al.* (1987) showed that schizophrenic patients do not experience comparable rates of recurrence or exacerbation of their illnesses—only one of 22 schizophrenic women (this group having a total of 29 episodes of childbirth) was admitted during the period of the study. A review (unpublished) of consecutive admissions (1987–8) to the Mother and Baby Unit at the Bethlem Royal Hospital, London, on the other hand, showed that 12 out of 57 mothers admitted during a 22 month period had a definite history of schizophrenia. Thus, in London, the availability of a specialized mother and baby unit greatly increased the admission rate for schizophrenic women in contrast with Edinburgh which did not have such a unit. The figures cannot, however, be directly compared because the Bethlem unit does not have a defined catchment area and it is not known how many schizophrenic mothers from the hospital's local catchment area were *not* admitted to the unit during 1987–8. Most of these admissions were not because the mother had experienced a flare-up of her illness, but rather to assess her motivation and competence safely to care for the newborn child, while at the same time her own needs for treatment were met. Such assessments are undertaken by the clinical team of the in-patient unit at Bethlem and final recommendations are made after close liaison with health professionals in the community and with representatives of social services who may be required to take steps to protect the child. Methods are being developed to try to assess parenting behaviour in a reliable way and thus to provide sounder methods than are currently available for the prediction of eventual 'safe' parenting. Nurses at the clinical unit at Bethlem have played a major part in the development of those assessment procedures (Thiels and Kumar 1987; Kumar 1986). It may well be the case that for many such mothers, in particular single women with chronic schizophrenic illnesses, admission is inappropriate and counter-therapeutic. At Bethlem over 90 per cent of such assessments eventually result in long-term fostering or adoption of the child away from the mother. It might be better to carry out assessments on a day attendance basis, with the infant fostered from the outset, but spending a limited time each week with its mother attending on a day basis at a unit with the skills and re-sources to provide an opinion on the mother's motivation and competence as a parent both now and in the future.

Some suggestions for research

The conclusion has to be that we have no substantive information about whether it really is better to look after mentally ill mothers with or without their young infants. Intuitively, it seems cruel to separate a mother from her baby, and nursing together, although more expensive, seems to work well without significant risk of harm to the child during the acute illness. Margison and Brockington (1982) document the infrequency of such incidents in one unit but, of course, one catastrophe is one too many. On the other hand, does anyone know what happens after separated, recovering mothers are reunited with their infants? It seems extraordinary that we know so little—why is this? It cannot be because the subject is not newsworthy, or that it is trivial in clinical or in numerical terms or in the repercussions the illness causes when a mother has to be admitted soon after giving birth.

The diversity of services that has developed in the UK can be turned to advantage, because it should enable one to select differing models of clinical care and then, having tried to match patients on given clinical and other characteristics, study outcome. A specialized mother and baby unit, a hospital which admits mothers and babies into a general ward setting, a day hospital, a hospital which admits mothers without babies and, finally, a service which aims to manage mothers at home are all possible candidates for a basis for such a survey of outcome. The greatest difficulty will arise in finding a service in the UK which routinely separates mothers from infants; going abroad to find such a service will naturally introduce possible error but there may be no alternative. The choice of outcome measures will depend upon the delay between the mother's illness and the time of follow-up. It may be useful to include both short- and long-term follow-up, perhaps combining a prospective investigation with a follow-up study of cohorts of women admitted to different kinds of facilities some time previously (e.g., 5–10 years ago).

Some of the most urgent questions have been raised in the previous sections and no doubt there are many others. Groups of women which have not been considered in this review include mothers who misuse drugs and alcohol, those with epilepsy, those with moderate to severe mental impairment, and those with personality disorders that are marked by impulsive and violent behaviour patterns. The methods are mostly available to tackle some of the questions.

Clinical guidelines

1. *Plan ahead*

Women with histories of psychiatric disorder who become pregnant are 'at risk' in a number of different ways. The likelihood of recurrence of affective psychosis after childbirth (up to 50 per cent in some studies) is high enough to warrant reserving a bed during the time of greatest risk, which is the first two weeks. Anticipatory provisions can thus be made to admit the baby with the mother, if there is no specialized unit available. The GP, health visitor, and community midwives should be alerted to the possibility of recurrence and effective liaison established with psychiatric services. Schizophrenic mothers, especially if they have a chronic illness and if they are single and lack family support, can present clinical staff with a serious dilemma. How can one comply with the mother's wish to raise the baby and yet ensure that the child's needs will be safely and adequately met? Anticipatory meetings with social services are very helpful because a period of assessment of mother and baby in a safe environment may be needed before any decisions can be reached. Similar problems can arise with mothers who have severe personality disorders and those who abuse drugs.

2. *Respond quickly*

Often there is an urgent call from a GP for a domiciliary visit because a newly delivered mother is unexpectedly behaving strangely or is becoming disturbed at home. Post-partum psychosis is an acute psychiatric emergency and clinical concern is heightened because of the presence of the young infant. One should ascertain in advance where it is possible to admit mother and baby together if this is what is appropriate. The admitting staff should be warned of possible risks to the infant (e.g., if it is enmeshed in the mother's delusional system or if she is voicing suicidal and possibly infanticidal ideas). Mothers should not be left unsupervised at home while a prolonged search is started for a suitable mother and baby place. It is better to admit only the mother urgently, having made temporary arrangements for the baby's care, and then to reunite them later.

3. *Admission*

Having admitted mother and baby, a management plan should be decided with the nursing team. Experience of managing such cases is obviously desirable in deciding whether or not the mother should be compulsorily detained or, if not, whether she or others should temporarily be prevented from removing the infant from hospital ('place of safety order'). The level of nursing observation and the extent to which the mother is allowed to care for the baby in the initial stages of the illness require positive decisions and then planned reviews. In hospitals which do not have any separate facilities or staff with designated responsibilities for mothers and babies, it seems sensible to restrict admissions to one clinical team. Even if such admissions are sporadic, expertise and resources are gradually built up in this way.

4. *Medical treatment*

Until there are specific, possibly hormonally based therapies of proven efficacy the medical management of post-partum psychosis is determined by the predominant psychopathology. The clinical picture can be pleomorphic and it invites polypharmacy. ECT can be dramatically effective and it has the advantage that mothers can continue to breast-feed without concerns about the transfer of drugs via milk. It is probably safe to breast-feed while taking antidepressants and neuroleptics but there is little systematic research to support or refute such statements. Breast-feeding while taking lithium is contra-indicated.

References

Aston, A. and Thomas, L. (1987). Mother and baby psychiatric facilities in England and Wales 1985–6. *Annual Bulletin of the Marcé Society* (ed. F. Margison), pp. 3–12. Pegasus, London.

Baker, A. A., Morison, M., Game, J. A., and Thorpe, J. G. (1961). Admitting schizophrenic mothers with their babies. *Lancet*, ii, 237–9.

Bardon, D., Galser, Y .I. M., Prothero, D., and Weston, D. H. (1968). Mother and baby unit: psychiatric survey of 115 cases. *British Medical Journal*, 2, 755–8.

Brockington, I. F., Cernik, K. F., Schofield, E. M., Downing, A. R., Francis, A. F., and Keelan, C. (1981). Puerperal psychosis. *Archives of General Psychiatry*, 38, 829–33.

Brockington, I. F., Winokur, G., and Dean, C. (1982). Puerperal psychosis. In *Motherhood and mental illness* (ed. I. F. Brockington and R. Kumar), pp. 37–69. Academic Press, London.

Buist, A., Dennerstein, L., and Burrows, D. G. (1990). Review of a mother and baby unit in a psychiatric hospital. *Australian and New Zealand Journal of Psychiatry*, 24, 101–8.

Dean, C. and Kendell, R. E. (1981). The symptomatology of puerperal illness. *British Journal of Psychiatry*, 139, 128–33.

Fowler, D. B. and Brandon, R. E. (1965). A psychiatric mother and baby unit. *Lancet*, i, 160–1.

Glaser, Y. I. M. (1962). A unit for mothers and babies in a psychiatric hospital. *Journal of Child Psychology and Psychiatry*, 3, 53–60.

Grunebaum, H., Weiss, J. L., Cohler, B.J., Hartman, C. R., and Galland, D. H. (1975). *Mentally ill mothers and their children*. University of Chicago Press.

Hamilton, J. A. (1982). The identity of postpartum psychosis. In *Motherhood and mental illness* (ed. I. F. Brockington and R. Kumar), pp. 1–17. Academic Press, London.

Hemphill, R. E. (1952). Incidence and nature of puerperal psychiatric illness. *British Medical Journal*, ii, 1232–5.

Inwood, D. G. (ed.) (1985). *Postpartum psychiatric disorders*. American Psychiatric Association, Washington.

Kendell, R. E. (1985). Emotional and physical factors in the genesis of puerperal mental disorders. *Journal of Psychosomatic Research*, 29, 3–11.

Kendell, R. E., Wainwright, S., Harley, A., and Shannon, B. (1976). The influence of childbirth on psychiatric morbidity. *Psychological Medicine*, 6, 297–302.

Kendell, R. E., Rennie, D., Clarke, J. A., and Dean, C. (1981). The social and obstetric correlates of psychiatric admission in the puerperium. *Psychological Medicine*, 11, 341–50.

Kendell, R. E., Chalmers, J. C., and Platz, C. (1987). Epidemiology of puerperal psychosis. *British Journal of Psychiatry*, 150, 662–73.

Kumar, R. (1982). Neurotic disorders in childbearing women. In *Motherhood and mental illness* (ed. I. F. Brockington and R. Kumar), pp. 71–118. Academic Press, London.

Kumar, R. (1986). Post-partum mental illness: a method for assessment of mother–infant interaction. In *Current approaches to puerperal mental illness* (ed. J. L. Cox, R. Kumar, F. R. Margison, and L. J. Downey), pp. 45–55. Duphar, Southampton.

Kumar, R. (1990). Childbirth and mental illness. *Triangle*, 29(2/3), 73–82.

Kumar, R., Meltzer, E. S., Hepplewhite, R., and Stevenson, A. D. (1986). Admitting mentally ill mothers with their babies into psychiatric hospitals. *Bulletin of the Royal College of Psychiatrists*, 10, 169–72.

Lindsay, J. S. B. and Pollard, D. E. (1978). Mothers and children in hospital. *Australian and New Zealand Journal of Psychiatry*, 12, 245–53.

Luepker, E. T. (1972). Joint admission and evaluation of postpartum psychiatric patients and their infants. *Hospital and Community Psychiatry*, 23, 284–6.

Margison, F. and Brockington, I. F. (1982). Psychiatric mother and baby units. In *Motherhood and mental illness* (ed. I. F. Brockington and R. Kumar), pp. 223–7. Academic Press, London.

Masson, D., Fivaz, E., and Ciola, A. (1977). Expérience d'hospitalisation conjointe mère-enfant dans un centre de traitement psychiatrique de jour (Hôpital de Jour) pour adultes. *Archives Suisses de Neurologie, Neurochirurgie et Psychiatrie*, 120, 83–100.

Meltzer, E. S. and Kumar, R. (1985). Puerperal mental illness, clinical features and classification: a study of 142 mother and baby admissions. *British Journal of Psychiatry*, 147, 647–54.

Mester, R., Klein, H., and Lowental, U. (1975). Conjoint hospitalisation of mother and baby in post-partum syndrome—why and how. *Israeli Annals of Psychiatry and Related Disciplines*, 13, 124–36.

Murphy, H. B. M. (1982). *Comparative psychiatry*. Springer, Berlin.

Nott, P. M. (1982). Psychiatric illness following childbirth in Southampton: a case register study. *Psychological Medicine*, 12, 557–61.

O'Hara, M. and Zekoski, E. M. (1988). Postpartum depression: a comprehensive review. In *Motherhood and mental illness. 2. Causes and consequences* (ed. R. Kumar and I. F. Brockington), pp. 17–63. Wright, London.

Paffenberger, R. S. Jr (1964). Epidemiological aspects of postpartum mental illness. *British Journal of Prevention and Social Medicine*, 18, 189–95.

Polonio, P. and Figueiredo, M. (1955). On the structure of mental disorders associated with childbearing. *Monatsschrift der Psychiatrischen Neurologie*, 130, 304–18.

Prettyman, R. J. and Friedman, T. (1991). Care of women with puerperal psychiatric disorders in England and Wales. *British Medical Journal*, 302, 1245–6.

Racamier, P. C., Sens, C., and Carretier, L. (1961). La mère et l'enfant dans les psychoses du post-partum. *Evolution Psychiatrique*, 26, 525–70.

Schuurmans, M. J. (1966). The psychotic mother and the healthy infant on the psychiatric unit. *Perspectives in Psychiatric Care*, **4**, 22–6.

Spitzer, R., Endicott, I., and Robins, E. (1978). *Research diagnostic criteria for a selected group of functional disorders* (3rd edn). New York State Psychiatric Institute, New York.

Stein, G. (1982). The maternity blues. In *Motherhood and mental illness* (ed. I. F. Brockington and R. Kumar), pp. 119–54. Academic Press, London.

Stern, G. and Kruckman, L. (1983). Multidisciplinary perspectives on postpartum depression, an anthropological critique. *Social Science and Medicine*, **17**, 1027–41.

Stewart, D. R. (1989). Psychiatric admission of mentally ill mothers with their infants. *Canadian Journal of Psychiatry*, **34**, 34–8.

Thiels, C. and Kumar, R. (1987). Severe puerperal mental illness and disturbances of maternal behaviour. *Journal of Psychosomatic Obstetrics and Gynaecology*, **7**, 27–38.

Van der Walde, P. H., Meeks, D., Grunebaum, H. U., and Weiss, J. L. (1968). Joint admission of mothers and children to a state hospital. *Archives of General Psychiatry*, **18**, 706–11.

16

Maximum security hospitals: can we do without them?

PAMELA J. TAYLOR

Introduction

At any one time during 1990 and 1991 there were about 1700 people from England and Wales compulsorily detained in a maximum security hospital in order to receive psychiatric treatment. The NHS Act 1977 described such patients as 'requiring . . . special security on account of their dangerous, violent or criminal propensities', and the three hospitals concerned—Ashworth near Liverpool, Broadmoor in Berkshire, and Rampton in Nottinghamshire—are generally known as special hospitals. Scotland has the comparable State Hospital at Carstairs. Much of what follows will apply to that, and maximum secure hospitals elsewhere in the world, but my first hand experience and knowledge is of the English special hospitals.

The meaning of special hospital care

Physical security

The deliberate qualities of physical security that distinguish the special hospitals from other hospitals are their high perimeter walls or fences, and their heavy, if patchy use of internal physical security. The high walls of Broadmoor and Ashworth North make them look more like top-security prisons although, unlike these, even they are not equipped to prevent major break-ins, merely unwarranted departures. The principle of internal and external locks differing, and internal and external keys never being held simultaneously is similar to that in medium secure units. Additional, largely fortuitous security lies in the physical isolation of the hospitals—not from the communities local to the hospitals, but from the original communities of the patients.

Security by staff numbers

The qualities and quantity of staff are at least as crucial to security as walls and locks. It seems paradoxical that nurse–patient ratios in special hospitals are lower than in medium secure units, and even general psychiatry special care units. In the special hospitals there is, in part, compensation for this in shift distributions. Staff availability by day is similar to other hospital security settings, but there is a skeleton staff at night, almost turning the hospitals into prisons. Thus all patients are routinely locked in their rooms or dormitories at around 9.00 p.m. without the possibility of leaving, even for the lavatory. This practice is so familiar that it is regarded neither by most patients nor by staff as seclusion. Patients in distress can rarely be counselled. Improved nocturnal staffing is crucial for the future.

Security by training

The safe treatment of patients in security depends on a cluster of special skills which must be built on to a sound general psychiatric training. For many years this has been recognized in the training of forensic psychiatrists (JCHPT 1990). A period of training in a maximum secure hospital setting is mandatory for full specialist approval in forensic psychiatry. Other disciplines have followed such principles, and an important development is a post-basic course in forensic psychiatric nursing at Ashworth Hospital, recognized by the English National Board (ENB). The necessary skills include the management of physical aspects of security; the co-ordination of systematic, basic observations of both environment and patients; recognition of more subtle developments in ward atmosphere; recognition of sexual or aggressive transferences and counter-transferences, whether or not these are encouraged in treatment; the ability to tolerate long-term relationships with hostile and deviant people; the use of personal relationships as a holding measure (Cox 1986); and a ready knowledge of the use of chemotherapy, to balance other physical methods of containment as well as for specific treatment.

Security by selection

Up to 10 patients a year may be directed into special hospitals by the Home Office under the Criminal Procedures (Insanity) Act 1964.* In common with other health service facilities, however, the staff otherwise choose their patients, although a multidisciplinary hospital panel debates the assessing consultant's recommendation. Disagreement is rare. Rejection on the grounds that the patient or his or her circumstances would render the admission too dangerous is extremely unusual (e.g., the likelihood of a strongly financed and professional rescue operation).

Selection also operates internally. Some segregation of patients by such criteria as 'untested', highly disturbed, well controlled or pre-discharge occurs, but is similar to that practised in medium secure units. Seclusion, meaning enforced confinement of a patient alone in a room as a response to disturbed behaviour, is probably more widely used in special hospitals than elsewhere. Its efficacy anywhere has yet to be established. It is likely at best

---

* Now: Criminal Procedure (Insanity and Unfitness to Plead) Act 1991.

to be a facility for containing dangerous behaviour rather than a primary therapeutic tool (Gunn and Taylor 1992).

Gender segregation, still applying almost as extensively to staff as to patients, sets the special hospitals apart from other psychiatric hospitals. A male patient may be exceptionally or exclusively dangerous to women, or a female to men, but this is unusual. In practice, where mixing occurs—in workshops, at social events, and in one ward—dangerous behaviour has been rare and responsible behaviour has been fostered. Although male-female ratios (4:1 among patients) make full integration difficult, much more could and should be done. By far the most serious breaches of internal security in 1990 were in the context of the intensity of a single sex climate and disturbed male homosexual behaviour.

Political and administrative security

Broadmoor Hospital was originally under the management of the Home Office, passing to the Board of Control after the Criminal Justice Act 1948, then the Department of Health and Social Security (DHSS) after the Mental Health Act (MHA) 1959. Of the other hospitals only Rampton was briefly under Home Office management. Direct management by the DHSS, and later Department of Health (DOH), came under increasing criticism in several key reports from the Health Advisory Service (HAS) on Broadmoor in 1975 and 1988, the House of Commons Estimates Committee in 1968, and the Boynton Committee of Enquiry into Rampton Hospital (DHSS 1980). In 1989, a new Special Health Authority, the Special Hospitals' Service Authority (SHSA) was set up exclusively for the special hospitals, with a requirement to report regularly to the DOH. The Home Office retains control over the movement of many individual patients from each of the hospitals and 68 per cent currently subject to Home Office restrictions on discharge.

Any tabloid press reader will be aware that sometimes political factors may influence the management of individual patients, as well as the hospitals. In 1985, for example, there were front page features on the possible release by Mental Health Review Tribunal (MHRT) of a named patient, with details of the case. A government minister was quoted as saying that the patient's release 'would terrify me'. An enforced adjournment of the MHRT followed. The House of Lords (31.1.91) subsequently upheld the liberty of the press in this regard.

Psychiatric treatment as security

An admission to special hospital for treatment may only be to fulfil the rights of a person to receive, in safety, treatment appropriate to his or her condition. More often than not, however, the psychiatric disorder was a material factor in the dangerous behaviour precipitating admission. Both milieu therapy and specific treatments then become essential to security.

Do special hospitals provide a unique service?

No single aspect of special hospital care is unique. The more striking aspects of physical security can be seen in prisons and the lesser aspects elsewhere in the health service. Treatments are broadly the same as elsewhere in the NHS, although few centres provide the expertise and length of commitment in psychotherapy as it is provided at its best in the special hospitals (Cox and Theilgaard 1987). It is the combination of treatment and high security that is characteristic of maximum security hospitals.

Maximum security hospitals are not unique to England. Scotland has its State Hospital, Eire's Dundrum preceded Broadmoor, and the US and many continental and commonwealth countries have them.

The demand for care

Historical perspective

Britain notionally coped without maximum security hospitals until 1863. Allderidge (1979) described centuries of ambivalence on the part of society and its service providers to its mentally ailing folk, particularly the chronically ill and the behaviourally frightening. Parker (1985) continued the story in relation to secure provision. Themes familiar to modern practice emerge. The county asylums of the early nineteenth century were reluctant to receive insane prisoners because the cost would fall on the parishes, whereas prisoners in jails were paid for by the counties. Other complaints were that the hospitals were full, or that the security required would infringe the liberty of the other patients, but some simply said that it was 'highly objectionable' to keep 'such persons' in a 'general lunatic hospital'. The prisons were no more enthusiastic about coping with the doubly disabled— the 'mad and bad'. Broadmoor Hospital was finally completed only following special legislation nearly 60 years after its first recommendation by a parliamentary select committee. This proved a precedent for procrastination in the delivery of services to mentally abnormal offenders. In December 1990, in response to the heavy criticism of the Prison Medical Service and of the NHS in not making appropriate treatment available for the antisocial and poorly socialized mentally ill, the Department of Health announced a new national working group (DoH, Home Office 1992).

Recent trends in referrals and bed occupancy

For admission to a special hospital, each patient must be compulsorily detainable under the 1983 Mental Health Act *and* pose a grave and immediate danger to others. Arguments about detainability are as many and varied as elsewhere in the NHS, and focus especially on the treatability of psychopathic disorder or mental impairment. In relation to dangerousness the factors listed for consideration include evidence of unprovoked or random physical or sexual assaults on members of the public; threatening psychotic symptoms which incorporate specific people; arson; the use of poison or weapons to harm others; sadistic behaviour; and hostage-taking. Great emphasis is placed on restricting admission to those who could not possibly be managed in any other hospital setting.

In 1972, there was a peak of 2300 patients in the special hospitals. Residency declined, to a low of 1627 in 1985, but has been rising again since, to a little above 1700. The fluctuations coincided with the opening of medium secure beds in the late 1970s and their reaching near current capacity of around 600 places in the mid-1980s. Whether the association is cause, effect or merely chance is difficult to say. The referral rate has fluctuated only slightly, the annual average for the five years between 1984 and 1988 being 314, with an acceptance rate of just over two-thirds.

Future demands and unmet need

Although past trends give a reasonable indicator of likely short-term future need, many extraneous factors potentially limit accurate longer term calculation. A breakthrough in treatment, although seeming unlikely, might occur, relieving patients with treatment-resistant schizophrenia, who constitute the majority of the population. If medium secure beds were provided to recommended levels, substantial numbers of transfers could be effected.

Other factors suggest a possibly increased call for special hospital beds. A conservative projection for long-stay hospital bed provision for mental illness alone is 50 per 100 000 (Robertson 1981, endorsed by Wing 1990). Special hospitals also provide for the personality disordered and the mentally and severely mentally impaired, but just 3.5 per 100 000 population in total. As ordinary psychiatric hospital closure continues the demand for special hospital beds will probably increase. Already, referrals are turned away on grounds that the only reason for admission would be the lack of more appropriate hospital provision, but if patients cannot get hospital treatment their dangerousness may escalate. Serious violence is usually a feature of well-established illness (Taylor in press). Delays in discharge from special hospitals because of the lack of appropriate alternatives have remained disturbingly high throughout the last decade (Dell 1980; Taylor *et al.* 1991).

Long-standing concern about the substantial and probably growing number of mentally disordered people inappropriately placed in British jails has been underscored by the escalating suicide rates (Dooley 1990; HM Chief Inspector of Prisons 1990). The problem is particularly acute among the largely untried remand populations, where up to 10 per cent are likely to have a psychotic illness (e.g., Taylor and Gunn 1984; Coid 1988). Serious personality disorders, drug and alcohol abuse, and sexual deviancy are even more common. Some of the sick people are filtered out of the remand prisons into treatment, but others may be denied by a diagnostic sleight of hand (Coid 1988). A 5 per cent sample of all sentenced prisoners in England and Wales (Gunn *et al.* 1991) found that about 2.5 per cent had a frank psychotic illness. Many more people needed assessment or treatment. It is difficult to estimate not only what level of security would be needed for these men and women, but also what level would be acceptable to the Home Office, which would have to sanction transfer. As many as 400 places could be required in the special hospitals in addition to similar numbers in medium secure and in open hospitals if the need were to be fully met.

Evidence of real need

Relationship between dangerous behaviour and mental disorder

The convicted criminal may develop co-incidental mental disorder, but the crime and the disorder may be linked. Monahan and Steadman (1983) provided the most comprehensive review of the literature to that time, high-lighting problems with the data but finding that, overall, there did seem to be a greater than chance association between mental disorder and crime, and Wessely and Taylor (1991) have extended the discussion. Swanson *et al.* (1990) may have come closer than any to identifying true crime and true disorder in the general population, although their work has been criticized on the grounds that the base rate of violence for the general population seems to be low and the apparently higher levels among some mentally disordered groups is thus perhaps exaggerated. They focused on violence in a substantial community survey in three USA states. Familiar associations with violence emerged, including youth, male gender, and low socioeconomic status, but it was also established that more than half of the sample met DSM-III-R criteria for one or more psychiatric disorders. Subjects with alcohol or drug problems were more than twice as likely to report violent behaviour as those with schizophrenia, but nearly 13 per cent of the latter reported violence at some time in one year, still about three to four times the rate of someone without disorder at all.

Violence in Swanson *et al.* (1990) was defined by personality ratings, but that serious violence does occur is undoubted. Hafner and Boker (1973), for example, calculated the risk of homicide by people with schizophrenia as about 5 in 10 000, and by people with an affective psychosis as ten times less likely. Among people with other disorders they placed the risks as very similar to those in the general population. Hafner and Boker (1973) and Taylor (1985) have also demonstrated strong links between the symptoms of psychosis and dangerous behaviour.

Is institutional care effective in containing danger?

The evidence that high security incarceration of violent, mentally disordered people is necessary or effective has never been sought. A few studies describe the violence of patients prior to entry to secure facilities of various kinds, and some describe the behaviour of entirely different cohorts once they have left. One or two observe changes in the short term after admission to intensive care. There are none that follow the careers of violent mentally disordered offenders from the point of entry to a secure hospital, through the hospital, and beyond. Violent behaviour, suicide, and even homicide occur in security hospitals. Some patients, therefore, may be regarded as continuing to require high levels of specialist care and control, but who are they? At the other extreme, it is not clear that when a previous act of violence is not repeated in the specialist facility this would indicate any success on the part of the environment. Among mentally normal and abnormal alike, some of the most serious violence, especially homicide, has an extremely low base rate.

Evidence that a particular secure facility is necessary to the containment of danger

Adequate tests of real need and real efficacy in provision are hard to devise. Perhaps the best hope lies in improving techniques for defining the circumstances of risk, and calculating as a consequence the minimum sufficient requirements to reduce that risk to acceptable levels. Calculations of risk must include the time exposed to the chances of acting violently. It must also mean improvement in the understanding of the relationship between violence and disorder in individual cases.

Establishment of true independence between violence and disorder in any given case would be useful, because the goals would be correspondingly simple: the provision of a safe environment in which effective treatment could be delivered for as long as that treatment remained necessary. There would be no expectation that treatment would have an overall influence on the individual's violent or antisocial behaviour. Conversely, establishment of a clear relationship between disorder and offending would make for relatively straightforward psychiatric judgements. If it could be shown that dangerous behaviour was, for example, driven by delusions, then three markers would indicate the relative need on psychiatric grounds for maximum secure placement as an adjunct to therapy: (1) the availability of effective specific treatment; (2) the rapidity with which it is likely to act; and (3) the chances of maintaining the patient on effective treatment once established. In reality, few patients fall neatly into any of these parameters, and safety calculations are more complex. A persecutory delusion, for example, may clearly provoke serious violence but the patient only requires maximum security if certain other factors prevail. If, for example, only one named person is threatened, and then only if regularly in close proximity to the patient, the resolution of risk is unlikely to require a special hospital, but when the threat is broad (e.g., to all brown-eyed women) and the patient's sense of personal boundary consistently poor, high security may be essential.

Patients currently in special hospitals

Few psychiatric disorders are not represented in the special hospitals. According to the 1983 Mental Health Act classification, 76 per cent of the men and 63 per cent of the women were mentally ill (almost all with schizophrenia) at the time of the 1990 census (Taylor *et al.* 1991). Referral rates for illness increased from 145 in 1984 to 192 in 1988 (DOH 1989). The mentally impaired, and the severely mentally impaired, form a diminishing group—only 200 in 1990. A fairly constant 25 per cent of special hospital patients have been classified as having psychopathic disorder—340 men and 108 women in 1990. Referral rates of the personality disordered remain steady at between 80 and 90 per year. Men outnumber women, and there are more younger than older people. At the time of the census, however, 95 people were over the age of 60. Maximum secure psychogeriatric care is an anomaly in most cases, although poisoning requires little strength and one 80-year-old continues to exhibit such prowess.

Most patients in special hospitals have been convicted of a criminal offence, but all have behaved dangerously: 68 per cent are under Home Office restrictions on discharge, a higher court having decided that in each case the public could be at risk of serious harm.

Do special hospital patients require special facilities?

Claims that the threshold for admission to special hospitals on security grounds is too low are rare; more often referring agencies regard it as too high. A serious problem, however, is that patients outstay their maximum secure need. At the time of the 1990 census just 42 per cent were said by their consultant psychiatrist to require either maximum perimeter security, maximum internal security, or both. Evidence of transfer delays has been accumulating for over 10 years (Dell 1980; Taylor *et al.* 1991). For nearly 20 per cent of the total, clinicians inside and outside the special hospitals had agreed a transfer plan, but did not have the facility to effect it.

It is ironic that there is more dispute about the appropriateness of the hospital element of the care in special hospitals than its security. This is mainly around people classified as having psychopathic disorder. In Dell and Robertson's (1988) study of men in Broadmoor, 96 of 127 psychotic men were said by their consultants to be unfit for transfer or discharge. For 61 per cent of these, treatment had failed to bring the symptoms under control. These patients were probably incurable but, as treatment did appear to limit harmful actions or some subjective distress, no one questioned the appropriateness of their continued detention in hospital. By contrast, only 38 of the 106 patients with psychopathic disorders were thought to need treatment, but this reflected some contradictions. Although it was thought that medication had only a small role, for over 90 per cent of cases milieu therapy was said to be important, and for over 90 per cent of cases under the age of 30 (68 per cent of the total) psychotherapy was said to be important. Two-thirds of the men, however, were receiving no specific treatment.

One or two psychiatrists and others (Chiswick 1987; Grounds 1987; Mawson 1983; Dell and Robertson 1988) have heavily criticized the committal of people with psychopathic disorder to hospital. They adopt a justice model and express concern that offenders committed to hospital on grounds of psychopathic disorder spend longer in custodial confinement than they would have had a hospital bed not been on offer. There is no evidence for this. A more generally accepted concern is that if patients are to be forced to accept treatment, there must be a reasonable prospect of its improving their lot in some way. These principles of 'treatability'—creating improvement or preventing deterioration—are enshrined in the Mental Health Act 1983 for these patients; however, there often is confusion between 'treatability' and 'curability'. Furthermore, either depend as much on the treatment available as on the patient. If a man entering a hospital with septicaemia dies because supplies of the appropriate antibiotic were insufficient, that does not mean that he is not treatable. Similarly, if a person with psychopathic disorder is not supplied with adequate treatment for his or her specific problems, then he or she is unlikely to change significantly. The cost of intensive and sustained psychological treatments is relatively high—but this may not be the only bar to its provision. There is too the problem of the effect of the patient on the potential therapist—the problem of the strength and variety of the counter-transference (Dubin 1989; Maier 1990). The special hospitals are the only health service facility in Britain even to attempt substantial provision for people with personality disorders and dangerous behaviour. Even so, there are only two units of 25 beds each within the special hospitals explicitly for personality disordered men, of whom there are 340, and none for women of whom there are 108. Some individual treatment on an *ad hoc* basis supplements this, but a radical review is required.

The appropriateness of the large institution

It is unlikely that there will be a time in the foreseeable future when maximum secure hospital care is no longer necessary. Equally, it is unlikely that more than a minority of patients will need it. There remains the issue, however, as to whether it is best provided in a small number of large institutions, or whether there are better models for delivery.

Large institutions have, or can have, several advantages for patients. It is possible, for example, to cluster substantial expertise and resource in large institutions that would not be cost effective in smaller ones. This particularly applies to the educational and occupational facilities currently in the special hospitals, unparalleled in other hospitals or in prisons. Wing and Brown (1970) have demonstrated the importance of well-designed activity for patients. In hospital, the severity of negative symptoms of disorder was shown to be related to the length of time doing nothing, but rapid return to 'the community' proved to be of no advantage without stimulation and activity (Brown *et al.* 1966), the average amount of time spent doing nothing by patients who were unemployed being higher than that of long-term in-patients in the comparison hospital. There can be highly therapeutic qualities in the 'good asylum' (Wing 1990).

Another potential advantage for the large institution is that size may permit considerable internal freedom of movement, which is both a humanitarian consideration and a safety factor when people require long-term detention. Close confinement can be a trigger to violence, and both people with schizophrenia (Horowitz *et al.* 1964) and violent men (Kinzel 1970) have been shown to have greater 'body-buffer zones' than normal people. The present special hospital population is illustrative of need. The average length of stay is 8.5 years, but nearly 20 per cent of patients have been resident for more than 15 years. Many of these patients pose relatively specific high risks—perhaps to children, perhaps to people with whom they have had intense relationships, but they pose low risks to more neutral adult figures. Of the 702 patients in the 1990 special hospital census rated as still requiring maximum security, nearly 60 per cent (415) were said to need maximum perimeter security alone. The critical size for cost effective provision of space and facilities to make years of detention not only tolerable, but minimally damaging and safe, is unknown.

The disadvantages of the large, geographically isolated institutions must not, however, be minimized. There has been no shortage of scandals and enquiries in ordinary psychiatric hospitals (Martin 1984) and the problems tend to get magnified in maximum security hospitals. Almost every country that has a maximum security hospital has had a major national scandal and enquiry related to the institutionalism, brutality, and more usually both, that have arisen (e.g., England: DHSS 1980; Norway: Davis 1980; Canada: Hucker 1986; New Zealand: Mason *et al.* 1988). There is a danger that staff become as marginalized and institutionalized as the patients, if not more so. The solutions lie in high standards of training, solid links with professional organizations, local university links (Knesper 1978), and strong local and catchment area health service links, all currently under development.

The most relevant points of liaison between the maximum security hospitals and other health services are the Regional Secure Units. If these liaisons work well, would it be feasible to develop local maximum secure services? Calculating on present levels of demand, the resultant units would not be so very small. The smallest health region in the country is East Anglia. Its population of about two million would dictate 70 maximum secure beds. It already has approximately 30 medium secure beds for its mentally ill patients, and is developing a security service for the mentally handicapped. Would it be possible to staff and resource a unit not much in excess of 100 beds for all secure hospital needs? Would it be possible to repeat proportionately larger structures in each of the other 13 health regions and in Wales? Some parts of the world have populations that would not sustain units even of this size—for example, Western Australia or Queensland—and yet are now committed to combining a full range of forensic psychiatry provision on one site. Should the UK wait and watch these developments or move in these directions anyway?

In the final analysis, decisions are likely to be driven by resource. Well-resourced spacious hospitals for smaller numbers would be more costly. Furthermore, the Health Regions of England and Wales have a very poor record in providing services for mentally abnormal offenders, even when special earmarked funds are made available (Snowden 1987). In 1991, 16 years after Butler negotiated this funding to underpin the recommendations in his committee's report (Home Office, DHSS 1975) only 600 or so of the 1000 secure beds agreed are built and operational and the provision of long-term medium security is an almost wholly unmet need.

Special (maximum secure) hospitals: the clinical issues

1. Mental disorder is probably more commonly associated with dangerous actions than would be expected by chance, with direct links between the disorder and the actions demonstrable in many cases.

2. A minority of these mentally disordered people need treatment in a maximum secure hospital bed. Combining evidence that some patients in special hospitals no longer require maximum security, with evidence of unmet needs among prisoners, the provision of 35 maximum secure beds per million population seems likely to hold good for the foreseeable future.

3. Some of the clinical dilemmas about precisely which patients will benefit from treatment can be resolved only through improvement in the range and availability of treatment and the evaluation of the impact of such change.

4. The principal unresolved dilemma is whether, in the longer term, maximum secure hospital care should be provided in just three hospitals of 500 beds or more each. The disadvantages of patient dislocation from their community of origin may be outweighed by the relative freedom of movement and the range of resources that can be provided cost effectively in large hospitals. The problems created by District or Regional Health Authority discrimination against relatively small numbers of economically and socially disadvantaged men and women could be offset by central management of funding, but this need not mean large 'central' hospitals.

5. Although these dilemmas are debated widely in health and penal services, the special hospitals must deliver improved treatment and care, without compromising longer term decisions about change. Good treatment is not incompatible with security, indeed, it is essential to it. Some of the changes already underway to ensure an appropriate balance between treatment and security include:

(a) Replacement or upgrading of deteriorated buildings with modern facilities.
(b) The discarding of prison-style uniforms.
(c) Greater emphasis on specific treatments and on individuality, with individual treatment plans for all patients.
(d) Conscious efforts to decrease the institutionalizing effect of the large hospitals. These include the setting up of rehabilitation directorates in each hospital and an adjustment of staffing to extend the length and variety of the patient's day, and to ensure that hospital conditions prevail at night.
(e) Related efforts to decrease the institutionalizing effect on staff, with encouragement to increase professional and academic links, to develop training and to participate in research. Establishment of catchment areas for the special hospitals has facilitated other health service liaison for all disciplines.
(f) Efforts to help others understand the extent of the problems encountered by staff within the special hospitals, including actual violence in some cases, and extreme prolonged hostile transferences from patients in many.
(g) Greater sensitivity to the real, sometimes terrible problems that must still occasionally arise in institutions primarily for violent people, with the establishment of satisfactory complaints procedures for patients and staff alike, including the recourse to open enquiry when complaints are serious.

Acknowledgements

My thanks to Charles Kaye for commentary on this paper and to Denise Formosa for preparing the manuscript.

References

Allderidge, P. (1979). Hospitals, madhouses and asylums: cycles in the care of the insane. *British Journal of Psychiatry*, **134**, 821–34.

Brown, G. W., Bone, M., Dalison, B. M., and Wing, J. K. (1966). *Schizophrenia and social care*. Oxford University Press.

Chiswick, D. (1987). Managing psychopathic offenders: a problem that will not go away. *British Medical Journal*, **295**, 159–60.

Coid, J. (1988). Mentally abnormal prisoners on remand. *British Medical Journal*, **296**, 1779–84.

Cox, M. (1986). The 'holding function' of dynamic psychotherapy in a custodial setting: a review. *Journal of the Royal Society of Medicine*, **79**, 162–4.

Cox, M. and Theilgaard, A. (1987). *Mutative metaphors in psychotherapy*. Tavistock, London.

Davis, D. R. (1980). Report on Norway's Rampton. *Mindout, December*, p. 6.

Dell, S. (1980). Transfer of special hospital patients to the NHS. *British Journal of Psychiatry*, **136**, 222–34.

Dell, S. and Robertson, G. (1988). *Sentenced to hospital*. Institute of Psychiatry, Maudsley Monograph No. 32. Oxford University Press.

DHSS (Department of Health and Social Security) (1980). *Report of the review of Rampton Hospital*. HMSO, London.

DoH (Department of Health) (1989). *Special hospitals patient statistics*. Unpublished document obtainable from Department of Health or Special Hospital Service Authority, Charles House, 375 Kensington High Street, London W14 8QH, UK.

DoH (Department of Health), Home Office (1992). *Review of health and social services for mentally disordered offenders and others requiring similar services* (The Reed Report). Department of Health, London.

Dooley, E. (1990). Prison suicide in England and Wales 1972–87. *British Journal of Psychiatry*, **156**, 40–5.

Dubin, W. R. (1989). The role of fantasies, counter-transference, and psychological defenses in patient violence. *Hospital and Community Psychiatry*, **40**, 1280–3.

Grounds, A. T. (1987). Detention of 'psychopathic disorder' patients in special hospitals: critical issues. *British Journal of Psychiatry*, **151**, 474–8.

Gunn, J. and Taylor, P. J. (ed.) (1992). *Forensic psychiatry: clinical, legal and ethical issues*. Butterworth–Heinemann, Oxford.

Gunn, J., Maden, A., and Swinton, M. (1991). *Mentally disordered prisoners*. Home Office, London.

Hafner, H. and Boker, W. (1973). *Crimes of violence by mentally abnormal offenders* (trans. Helen Marshall 1982). Cambridge University Press.

Her Majesty's Chief Inspector of Prisons for England and Wales (1990). Report of a review of suicide and self harm in prison service establishments in England and Wales. HMSO, London. Cm1383.

Home Office/DHSS (1975). *Report of the committee on mentally abnormal offenders*. HMSO, London. Cmnd 6244.

Horowitz, M. J., Duff, D. F., and Stratton, L. O. (1964). Body-buffer zone. *Archives of General Psychiatry*, **11**, 651–6.

Hucker, S. (1986). *Oak Ridge: a review and an alternative*. Ontario Ministry of Health, Toronto.

JCHPT (Joint Committee on Higher Psychiatric Training) (1990). *Handbook*. Obtainable from the Royal College of Psychiatrists, London SW1X 8PG.

Kinzel, A. F. (1970). Body-buffer zone in violent prisoners. *American Journal of Psychiatry*, **127**, 59–64.

Knesper, D. J. (1978). Psychiatric manpower for state mental hospitals. *Archives of General Psychiatry*, **35**, 19–24.

Maier, G. J. (1990). Psychopathic disorders: beyond counter-transference. *Current Opinion in Psychiatry*, **3**, 766–9.

Martin, J. P. (1984). *Hospitals in trouble*. Blackwell, Oxford.

Mason, K. H., Ryan, A. B., and Bennett, M. R. (1988). *Report of the Committee of Enquiry into proceedings in certain psychiatric hospitals in relation to admissions, discharge, or release on leave of certain classes of patients*. New Zealand Ministry of Health, Wellington.

Mawson, D. (1983). 'Psychopaths' in special hospitals. *Bulletin of the Royal College of Psychiatrists*, **7**, 178–81.

Monahan, J. and Steadman, H. J. (1983). Crime and mental disorder: an epidemiological approach. In *Crime and justice: an annual review of research*, Vol. 4 (ed. M. Tonry and N. Morris), pp. 145–89. University of Chicago Press.

Parker, E. (1985). The development of secure provision. In *Secure provision* (ed. L. Gostin), pp. 15–65. Tavistock, London.

Robertson, G. (1981). *The provision of in-patient facilities for the mentally ill. A paper to assist NHS planners*. DHSS, London (unpublished).

Snowden, P. R. (1987). Regional secure units: arriving but under threat. *British Medical Journal*, **294**, 1310–11.

Swanson, J. W., Holzer, C. E., Ganju, V. K., and Jono, R. T. (1990). Violence and psychiatric disorder in the community: Evidence from epidemiologic catchment area survey. *Hospital and Community Psychiatry*, **41**, 761–70.

Taylor, P. J. (1985). Motives for offending among violent and psychotic men. *British Journal of Psychiatry*, **147**, 491–8.

Taylor, P. J. (in press). Schizophrenia and crime: distinctive patterns in association. In *Mental disorder and crime* (ed. S. Hodgins). Sage, New Park, Calif.

Taylor, P. J. and Gunn, J. (1984). Violence and psychosis. 1. Risk of violence among psychotic men. *British Medical Journal*, **288**, 1945–9.

Taylor, P. J., Butwell, M., Dacey, R., and Kaye, C. (1991). Within maximum security hospitals: A survey of need. SHSA, London (unpublished).

Wessely, S. and Taylor, P. J. (1991). Madness and crime: criminology versus psychiatry. *Criminal Behaviour and Mental Health*, **1**, 193–228.

Wing, J. K. (1990). The functions of asylum. *British Journal of Psychiatry*, **157**, 822–7.

Wing, J. K. and Brown, G. W. (1970). *Institutionalism and schizophrenia*. Cambridge University Press.

17

Survivors of disaster: how can they best be helped?

PETER E. HODGKINSON

Introduction

Disasters vary on a continuum from mass to personal, from the Zeebrugge or Hillsborough tragedies to the everyday catastrophes of car crashes and house fires. Both ends of the continuum possess the same characteristic—a sudden, traumatic event tears the normal framework of life asunder.

The nature of the problem

The problems of disaster survivors fall into two categories: first, is the nature of their experiences; secondly, is the nature of their attitudes to help.

The experiences of disaster survivors

Much recent professional interest has focused on post-traumatic stress disorder (PTSD) (APA 1987). As formulated in DSM-III-R, PTSD has three main components.

1. *Re-experience phenomena* These include recurrent and intrusive distressing recollections or dreams of the traumatic event, sudden acting or feeling as if the event were recurring, and intense psychological distress at exposure to events that symbolize or resemble an aspect of it. One symptom is required to qualify for a diagnosis of PTSD.

2. *Avoidance or numbing phenomena* Avoidance includes efforts to avoid thoughts or feelings associated with the trauma, or avoidance of activities or situations connected with it. Emotional numbing includes markedly diminished interest in significant activities, feelings of detachment or estrangement from others, a restricted range of affect, or a sense of foreshortened future. Three symptoms are required to qualify for a diagnosis of PTSD.

3. *Symptoms of increased arousal* These include difficulty in falling or staying asleep, irritability or outbursts of anger, difficulty in concentrating,

hypervigilance, exaggeraged startle response, and physical reactions upon exposure to events that symbolize or resemble the trauma. Two symptoms are required to qualify for a diagnosis of PTSD, and all symptoms must be present for one month.

Clinical experience suggests validity of the concept of PTSD, but it must be recognized that certain important features of post-traumatic reactions are omitted. People may experience post-traumatic reactions which are as equally disabling as PTSD, yet not qualify for the diagnosis. Guilt, or 'survivor guilt', (Lifton 1967), is not included. Evidence from the Zeebrugge disaster, for instance, suggests that guilt is an important component and determinant of types of reactions. Thus, guilt about what one did is related to avoidance symptomatology, whereas survivor guilt (feeling guilty for being alive) is related to intrusive symptomatology (Joseph 1991).

What percentage of people suffer PTSD after experiencing trauma? This question is not easy to answer, because not all of the studies of disaster have used PTSD as the criterion. In a study of a factory fire, 36 per cent of survivors suffered PTSD (Weiseath 1989), compared to 23 per cent in a study of those affected by terrorist bombings (Loughrey *et al.* 1988). In assessment of survivors of the Zeebrugge disaster for compensation purposes, 90 per cent were suffering from PTSD (Williams and Yule 1988). The appearance of PTSD may also be delayed, and it must be recognized as having a cyclical character, with sufferers swinging from periods of intense intrusive memories to periods of mental avoidance, where they appear to be little troubled.

Roughly averaging various studies (Hodgkinson and Stewart 1991), 45 per cent of survivors appear to be experiencing significant disturbance. This includes anxiety reactions and depression as well as PTSD—indeed, depression and PTSD often coexist. Thus, 34 per cent of survivors of terrorist bombing were suffering from depression as well as PTSD (Loughrey *et al.* 1988), and the majority of those seen for medico-legal assessment following the Zeebrugge disaster who had PTSD also were found to be depressed. There are several dangers here. First, PTSD itself may be mis-diagnosed as depression by the inexperienced clinician, particularly as many of the symptoms of emotional numbing are similar to depression. Altern-atively, depression may be correctly diagnosed, and the post-traumatic element again missed.

There are also survivors who are bereaved and relatives who are bereaved but did not experience the impact—the accident itself. Evidence suggests that death which is unexpected and violent creates problems for the resolu-tion of grief (Parkes and Weiss 1983). Certain components of sudden violent death are responsible for this enhancement.

(a) *Shock* The normal sense of shock may be compounded by disbelief and a sense of bizarre reversal—the deceased was involved in a normally safe or enjoyable activity, and suddenly perishes in horrifying circumstances.

(b) *Guilt*   Guilt, again a normal reaction, may be enhanced by actions taken or not taken during the impact; the nature of last partings (sometimes angry ones in ambivalent relationships), and an intense sense of 'unfinished business'.

(c) *Anger*   This may be enhanced by the fact that often in man-made disaster there is a justifiable focus for anger in the form of a person or persons. This anger may be channelled into litigation or other legal procedures.

(d) *Intervention of authority*   This can cause confusion and uncertainty, especially in relation to the restricting of information to relatives at a time they simply 'need to know'.

(e) *Aspects of the death*   In some disaster scenarios, no body may be recovered to provide a focus for grieving. In others, the body may be badly damaged and relatives are encouraged not to view it. This may create a sense of unreality of the death. Evidence suggests that those who view bodies show better adjustment in the long term (Singh and Raphael 1981; Hodgkinson in press), although they may be more distressed in the short term.

(f) *Multiple bereavement*   Those who are multiply bereaved are coping not only with the increased effects of several losses, but with the virtual eradication of social support networks, not only in terms of physical absence, but in terms of other relatives unavailable due to their own grief.

Disaster survivors' attitudes to help

Lifton (1967) was the first to draw attention to the 'nurturance conflicts' of disaster survivors, who are in truth a 'reluctant population' (Lindy and Green 1981). This suspicion of offers of help from outsiders, the fear that such offers may be false, and the reluctance to take them up, has several causes.

The first is related to the trauma itself. There is a sense of feeling weak, demeaned, and victimized. The world becomes a dangerous and untrustworthy place, the catastrophe signifying this. Offers of help are from this same, untrustworthy world.

The second reason is related to the nature of survivors themselves. They are simply a random selection of ordinary people, whose only reason for victimization is being in the wrong place at the wrong time. They often know nothing of emotional help or the professionals who offer it. They may have no real understanding of what their feelings represent, and if they do, have no idea what to do or where to go. Talking about feelings to a professional is profoundly stigmatizing. One elderly survivor of the Bradford fire when approached by a social worker exclaimed: 'Oh no . . . the Welfare . . . that it should come to this!'

What is trauma?

Trauma is not simply a collection of symptoms of PTSD. These symptoms are the outward indicators of an internal process.

Prior to the trauma, the individual is a person with a history which gives them a sense of identity. The trauma literally cuts across the person's life, plunging them into a whirlpool of unfamiliar feelings of terrifying intensity. Within this, the individual's sense of invulnerability may be gone, the world may now be a chaotic, dangerous, unpredictable place, and the individual may feel that the event has happened because he or she is a bad person. The person may now feel, and be seen by others, as totally different to the one who existed prior to the trauma. Recovery is therefore not about returning to 'how things were', but going forward to a new adjustment in which the person reappraises himself or herself, in terms of personal capability for reacting in certain ways, and the world itself. Only by doing this can the individual's past be meaningfully reintegrated.

What influences recovery?

Three dimensions influence recovery (Hodgkinson and Stewart 1991).

1. *Dimensions of the person*

The individual's past history of dealing with tragedy and loss, and their beliefs and attitudes about themselves and the world, will influence how they react. Studies of volunteer firefighters of the Australian Ash Wednesday bushfires (McFarlane 1987, 1988, 1989) show that neuroticism was correlated with the development of PTSD. This is problematic, of course, as the post-disaster level of neuroticism cannot be assumed to represent that existing pre-disaster. One feature that appears in McFarlane's and other studies (e.g., Davidson *et al.* 1985), however, is that a history of previous psychiatric disorder, or familial psychiatric disorder, predisposes to PTSD.

Following the Zeebrugge disaster, those who described themselves on initial interview as a 'worrier', as being less reliant on others in terms of not asking for help, or as unable to express feelings, were all suffering more intensely during the first year (Hodgkinson, in press). Those who had a set of attitudes indicating that they felt they should keep their feelings under control were suffering more two-and-a-half years later (Joseph 1991).

2. *Dimensions of the trauma*

The nature of the trauma will influence the nature of the stress. Difficulties will arise if the stressor is intense or severe, involves heat and noise or darkness, is sudden, unanticipated and uncontrollable, irregular, occurs in large chunks, and involves loss, such as bereavement, threat to life, personal injury, or exposure to death (Rachman 1980). Other important elements include the duration of the trauma, whether it was experienced alone or with others, whether the threats involved in the event were single or multiple, and whether there is a possibility that they may recur. Different levels of exposure to disaster stress are known to influence the prevalence of PTSD (Weiseath 1989).

Perceptions of the impact are important. Thus, those in the Zeebrugge disaster who felt helpless during the impact, thought they let themselves or others down, experienced guilt over actions or their absence, experienced 'survivor guilt', or saw their escape as due to others, rather than themselves, were faring worse two-and-a-half years later (Joseph 1991). Such perceptions may well, of course, be governed by stable characteristics of the individual.

3. *Dimensions of the recovery environment*

Key elements in the recovery environment will include isolation and levels of social support, with less social support leading to higher symptom levels (Joseph 1991). In addition, perceptions of the helpfulness of support (Singh and Raphael 1981), and ongoing stressors in the shape of further life events, some of which will be connected to the disaster and some of which will not (Joseph 1991), are also related to symptom levels. Further important issues will be cultural rituals for recovery, and community, society, and media attitudes towards the event.

What helps disaster survivors? Clinical guidelines

As yet, there is a relative lack of empirical studies in the area of therapeutic management. These guidelines are therefore largely based upon clinical experience.

1. *Formulation*

The goal for disaster survivors is to gain a sense of mastery over the experience, and to integrate it into their understanding of themselves and the world. This is primarily achieved through developing an understanding of what has happened, what is currently happening, and what is going to happen. Lifton (1987) describes this as 'formulation'. It is essentially a cognitive task. Hodgkinson and Stewart (1991) identify a 'hierarchy of formulation' through which survivors can be assisted:

(i) *Why did it happen?* Survivors need information as to how the accident happened, to make sense of the seeming senselessness.

(ii) *How did I escape?*   Survivors often need information about the events following the impact. Survivors of the Zeebrugge disaster were only able to make sense of their escape by reference to a plan of the ship, or returning to an identical ship. Relatives of the deceased will want to know where the loved one was sitting, etc., and what were they doing.

(iii) *Why did I escape?*   Being able to mentally work through, by opening for examination the personally ascribed reasons for escape may have some moderating influence on 'survivor guilt'.

(iv) *Why do I feel like this?*   Survivors need information about the range of feelings they may experience, so that they are able to place their feelings within a framework, rather than viewing them as indications of impending insanity which they alone are experiencing. Leaflets attempting to achieve this have been used in the aftermath of many of the major disasters in the UK after 1985. Following the Zeebrugge disaster, over half the survivors and just under half the bereaved found such a leaflet helpful. The remainder had either never read it or had mixed feelings. Only a small percentage of the bereaved found it harmful (Hodgkinson in press).

(v) *What does the way I feel now mean about me as a person?*   Survivors may need assistance in attempting to put their present reaction in the context of the people they were. If this is not achieved, they may develop intense self-criticism for their 'failure to cope'.

(vi) *What does all that I have been through mean about the way I understand life?*   Survivors need to be allowed to express their altered attitudes to life, such as 'life is unfair', and open them to examination, challenge, and appropriate modification.

Information is, as has been seen, the key to many of the aspects of formulation. Disaster survivors, however, have time-dependent needs, which can only be met within a certain time frame. Thus, a body can only be viewed before it is buried or cremated—afterwards, the choice is lost. This means that people must readily be given information, rather than having to ask, and should be given a genuine choice about access to certain things. A genuine choice, in the example of viewing a corpse, would not involve saying, 'I shouldn't see him if I were you . . .', but rather, 'Your husband's body is damaged in the following ways . . .'.

2.   Outreach

Outreach is a tool of 'crisis intervention'. We know that survivors will not readily come forward for help of their own accord. In disaster outreach, the survivor is visited, being given the choice to reject rather than accept the visit. Following the Zeebrugge disaster, overall, 85 per cent of victims accepted such a visit (Hodgkinson 1990). Seventy per cent of survivors and 62 per cent of bereaved reported this as either 'helpful' or 'very helpful'. A small percentage reported the visits as harmful, probably because they were at the avoidance stage of the PTSD cycle, and the visit disrupted their current way of coping.

The purpose of an outreach visit is to communicate information, offer a point of contact for further information, and to allow the victim to state needs and assess whether the outreach worker can help with these.

In order to achieve this, help has to be offered in certain ways. It must be *credible*, that is, seen to be of real use to make sense to the victim, and *acceptable*, presented in a way that the victim does not feel demeaned. It must also be *accessible*, in that clients must not have to argue their eligibility; *continued*, in terms of guaranteed availability for two years; and *terminable*. *Confidentiality* will also be an issue for clients.

3. *Debriefing*

Psychological debriefing (Mitchell 1983; Hodgkinson and Stewart 1991) is a specific tool of crisis intervention, which is of particular value with groups of people who have experienced traumatic events together. However, the format may be followed with an individual just as readily.

The overall aim of a group debriefing is to minimize the occurrence of unnecessary psychological suffering, but individual objectives include the following:

- full expression of impressions, reactions, and feelings;
- decrease in individual and group tension;
- promotion of cognitive organization through clear understanding of both events and reactions;
- mobilization of resources within and without the group, increasing support, solidarity, and cohesiveness;
- decrease in the sense of uniqueness/abnormality, i.e., normalization through sharing;
- preparation for experiences to come; and
- identification of avenues of further assistance if required.

The ideal timing of a debriefing is 48 to 72 hours after the event, but debriefings after this period of time may also be valuable. The maximum number of participants is 15. Two personnel are helpful, with one as an identified leader. Debriefing is not group therapy in the usual sense, it is a well directed and structured format, which has a strong educative element. It has seven phases.

(i) *Introductory phase* This is very important, the introduction and setting of rules to ensure confidentiality, etc., decreasing the likelihood of malfunction.

(ii) *Fact phase* In response to the question 'What happened?', each person briefly describes what happened; how they came upon the event; the time sequence, etc. The aim is to achieve a clear correct picture of the facts of the event for all.

(iii) *Thought phase* In this phase, participants focus on decisions and thought processes as the event unfolded. Clarifying these will often give the key to understanding later reactions such as guilt.

(iv) *Reaction phase* This is often the longest part of the debriefing. When people describe reasons behind decisions they will often talk about fear and helplessness, self-reproach, etc. It is this that establishes similarities and normality. The facilitation of the expression of feelings is a process of modelling for participants in how to deal with them with each other/ families/friends in the weeks to come.

(v) *Symptom phase* Here, participants describe emotional, cognitive or physical symptoms during the event, on returning home, in the meantime and currently.

(vi) *Teaching phase* In this phase the facilitator identifies normality by reference to research or material from other debriefings. People are given a framework within which to expect reactions.

(vii) *Re-entry phase* In this phase, the debriefing is summed-up and further needs identified, for example, for individual contact, or follow-ups to the debriefing.

Although the effectiveness of debriefing *per se* has not been tested, the principles behind it are central to treating combat stress. They have been shown to enhance return to duty and to reduce the risk of further problems (Solomon and Benbenishty 1986).

4. *Self-help*

Contact with other survivors is powerful in reducing the sense of isolation survivors experience. This may through a support group, or the production of something such as a newsletter on behalf of the whole survivor group. These groups are most helpful when they have some 'outsider' present, otherwise they run the risk of becoming isolated and introspective.

5. *Therapy*

Some survivors may find minimal crisis intervention sufficient to facilitate return in the direction of normal functioning. Others may need more intensive therapeutic help.

As was indicated earlier, the first step is correct differential diagnosis with regard to PTSD and associated problems versus depression and anxiety states. Medication may be considered—those who are suffering problems with sleep sometimes find that brief use of hypnotics is helpful. The impact of medication on PTSD is a largely unexplored area—antidepressants have been used to mixed effect (Davidson *et al.* 1987), but clearly may be of value when depression is present as a separate entity.

One of the aims of therapy is to start an effective processing of the trauma. Treatments of a cognitive or psychotherapeutic nature may be appropriate in this. Alternatively, symptomatic treatment of elements of PTSD may be appropriate, using behavioural techniques.

(i) *The traumatic event—memories* Processing must begin with the events themselves. Any treatment package must therefore begin with a full exploration of the trauma itself in immediate debriefing or later in psychotherapy or behavioural therapy. Often only by a step-by-step analysis of this person's thoughts and reactions during the event can its 'personal meaning' be uncovered.

There are some people for whom exploration of the trauma is impossible. To proceed along this tack would simply lose them from therapy. They must be enabled to manage feelings rather than explore them further. They may be taught distraction techniques to manage ruminative thinking, or eye movement desensitization may be used to modify intrusive imagery (Shapiro 1989). The therapeutic mechanisms of this technique, which reduces the disturbing potential of intrusive images, is uncertain. The subject holds the disturbing image in his or her mind while making a series of rapid eye movements.

(ii) *Cognitive appraisal and dysfunctional beliefs* As old belief systems are shattered, new belief systems arise which affect the way the event itself and the survivor's role in it, and the world in general, are viewed or appraised. Cognitive therapy can be employed to attempt to modify distorted thinking, automatic negative thoughts or irrational beliefs. The three symptoms of PTSD most amenable to cognitive restructuring are: (1) Increased anger and irritability; (2) detachment and withdrawal; and (3) guilt (Keane *et al.* 1985).

(iii) *Over-arousal* Relaxation (aided perhaps by biofeedback techniques) and systematic desensitization or imaginal flooding can be aimed at reducing arousal in general and that due to specific thoughts or images (Schindler 1980).

(iv) *Avoidance*   *In vivo* desensitization or flooding may be an important tool in allowing people to return to situations where the trauma occurred and achieve some mastery of their anxiety (Fairbank *et al.* 1981).

(v) *Emotional numbing*   Activity structuring may reduce both the depressive feelings and the withdrawal from normal activities. Marital counselling or family therapy may be used to tackle the relationship difficulties that the person's withdrawal and irritability creates.

References

APA (American Psychiatric Association) (1987). *Diagnostic and Statistical Manual of Mental Disorders* (3rd edn, revised). American Psychiatric Association, Washington DC.

Davidson, J., Swartz, M., Storck, M., Krishnan, R. R., and Hammett, E. (1985). A diagnostic and family study of posttraumatic stress disorder. *American Journal of Psychiatry*, **142**, 90–3.

Davidson, J., Walker, J. I., and Kilts, C. (1987). A pilot study of phenelzine in the treatment of PTSD. *British Journal of Psychiatry*, **150**, 252–5.

Fairbank, J. A., DeGood, D. E., and Jemkins, C. W. (1981). Behavioural treatment of a persistent posttraumatic startle response. *Journal of Behavior Therapy and Experimental Psychiatry*, **12**, 321–4.

Hodgkinson, P. E. (1990). The Zeebrugge disaster. III. Psychosocial care in the UK. *Disaster Management*, **2**, 131–4.

Hodgkinson, P. E. (in press). Patterns of recovery for the Zeebrugge disaster. In *Psychological aspects of disaster* (ed. D. Lane and A. Taylor), pp. 257–94. Professional Development Foundation and British Psychological Society Group in Counselling Psychology, London.

Hodgkinson, P. E. and Stewart, M. (1991). *Coping with catastrophe*. Routledge, London.

Joseph, S. (1991). Attributions and emotional processing in victims of major disaster. Doctoral thesis, University of London.

Keane, T. M., Fairbank, J. A., Caddell, J. M., Zimering, R. T., and Bender, M. E. (1985). A behavioural approach to assessing and treating post-traumatic stress disorder in Vietnam veterans. In *Trauma and its wake*, Vol. 1 (ed. C .R. Figley), Brunner Mazel, New York.

Lifton, R. J. (1967). *Death in life: Survivors of Hiroshima*. Simon & Schuster, New York.

Lindy, J. D. and Green, B. L. (1981). Survivors: Outreach to a reluctant population. *American Journal of Orthopsychiatry*, **51**, 468–78.

Loughrey, G. C., Bell, P., Kee, M., Roddy, R. J., and Curran, P. S. (1988). Post-traumatic stress disorder and civil violence in Northern Ireland. *British Journal of Psychiatry*, **153**, 554–60.

McFarlane, A. C. (1987). Life events and psychiatric disorder: The role of a natural disaster. *British Journal of Psychiatry*, **151**, 362–7.

McFarlane, A. C. (1988). The aetiology of post-traumatic stress disorders following a natural disaster. *British Journal of Psychiatry*, **152**, 116–21.

McFarlane, A. C. (1989). The aetiology of post-traumatic morbidity: Predisposing, precipitating and perpetuating factors. *British Journal of Psychiatry*, **154**, 221–8.

Mitchell, J. T. (1983). When disaster strikes—The critical incident stress debriefing process. *Journal of Emergency Medical Services*, **8**, 36–9.

Parkes, C. M. and Weiss, R. S. (1983). *Recovery from bereavement*. Basic Books, New York.

Rachman, S. (1980). Emotional processing. *Behaviour Research and Therapy*, **18**, 51–60.

Schindler, F. E. (1980). Treatment by systematic desensitization of a recurring nightmare of a real life trauma. *Journal of Behavior Therapy and Experimental Psychiatry*, **11**, 53–4.

Shapiro, F. (1989). Efficacy of the eye movement desensitization procedure in the treatment of traumatic memories. In *Psychological stress and adjustment in time of war and peace*. The 4th International Conference, Tel-Aviv, Israel.

Singh, B. and Raphael, B. (1981). Postdisaster morbidity of the bereaved. *Journal of Nervous & Mental Disease*, **169**, 203–12.

Solomon, Z. and Benbenishty, R. (1986). The role of proximity, immediacy and expectancy in frontline treatment of combat stress reaction among Israelis in the Lebanon war. *American Journal of Psychiatry*, **143**, 613–17.

Weiseath, L. (1989). A study of behavioural responses to industrial disaster. *Acta Psychiatrica Scandinavia*, **80**(Suppl. 355), 13–24.

Williams, R. M. and Yule, W. (1988). The assessment of 'nervous shock' in compensation litigation. In *Traumatic stress*. The 1st European Conference, Lincoln.

18

The life chart: historical curiosity or modern clinical tool?

MICHAEL SHARPE

What is a life chart?

According to the *Oxford English Dictionary*, a chart is a navigator's sea map that shows coast outlines, rocks, and shoals. Just as such a chart depicts the relationship in space of features of the sea relevant to the purpose of the sailor, a life chart illustrates the relationship in time of events and behaviours and illness in a person's life in a way that is useful to the psychiatrist.

Three forms of life chart may be distinguished. The first consists of a simple listing in parallel columns, of information about various aspects of a person's life. The second is a simple graphical representation of the pattern of recurrence of illness, in the form of a calendar. The third combines features of both of these by plotting the pattern of illness and its association with relevant factors from the person's life, in a graphical form against time (see Fig. 4). This type of chart has been used by several groups of research workers in recent years, but accounts of its use in clinical practice are few (see Sharpe 1990).

Although life charts have a long history in psychiatry, do they have anything to offer the busy modern clinician? In considering this question I will first review the origins of charts and consider whether they are more than historical curiosities. I will then discuss more recent developments in life charting and evaluate the potential role of charts as clinical tools in a modern psychiatric service. Finally, I will offer guidelines for their construction and use.

The history of life charts in psychiatry

Even though the chronologically ordered listing of clinical information has a long history in psychiatric practice, the first accounts of simple graphical representations of the course of illness were not published until the beginning of this century. The earliest examples are found in the writings of Kraepelin, who used such charts to illustrate regularities in the course of manic-depressive illness (Kraepelin 1913). At about the same time, Adolf Meyer introduced more complex charts that portrayed not only the temporal relationship of periods of illness, but also their relationship to other events and influences occurring over a person's lifetime (Meyer 1917/1952). He advocated the life history as the best way to understand an individual—the life chart being an ideal method. In 1917, Meyer wrote: 'No words of mine can give you a more graphic picture of the concreteness of what counts than the life chart'. Meyer's use of life charts reflected his psychobiological theory. The charts indicated the integration of the bodily organs into the whole person by displaying the weight curve of the organs in parallel down the centre of the chart, while important environmental influences and periods of illness were listed on each side.

Historical curiosity?

The historic charts of Kraepelin and Meyer could arguably be regarded as relics of a bygone age. Kraepelin's charts show only the natural history of illness without reference to life events and other factors that we now regard as important determinants of the course of illness. Meyer's original charts, with their speculative display of the weight of bodily organs, are complex and obscure to the modern clinician. Both obviously required much painstaking research for their construction. However, before dismissing the life chart as a curiosity it is necessary to review contemporary forms used in clinical research.

Modern charts

More recently, a form of life chart has been devised as a tool for clinical research. The modern form combines the simplicity of Kraepelin's charts in indicating the pattern of illness over time, with the illustration of diverse influences on the person and the course of their illness (whether biological, psychological or social) as in Meyer's charts (see Figs 5 and 6).

Manic-depressive disorder, with its recurrent nature, lends itself to charting. One group of researchers have developed the use of detailed charts in their study of this disorder (Squillace *et al.* 1984; Roy-Byrne *et al.* 1986; Post *et al.* 1988). Other investigators have used similar charts to study the course of schizophrenia (Harding *et al.* 1989) and drug addiction (Vaillant 1966), and to evaluate the long-term efficacy of psychiatric treatment (Schou 1973; Poynton *et al.* 1988).

Charts can be easily adapted to the aims of the investigator. An example is the use of a simple chart to examine the relationship between personal productivity and the course of manic-depressive disorder. Figure 4 was drawn to illustrate the correlation of mood and number of compositions

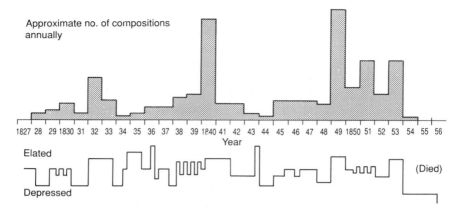

Fig. 4. Schumann's annual output of compositions and their relation to his documented mood swings [redrawn with permission from Trethowan (1977)].

completed by the composer Schumann during the course of his life (Trethowan 1977). Modern forms of life charts cannot be dismissed as mere historical relics. But do they have any role as a clinical tool?

Clinical tool?

The charts devised by modern research workers are clearly more relevant to the modern clinician's concerns than earlier versions. Do they have a role outside a specialized research setting? A simple chart takes an hour or more to create, and if further interviewing is required the time spent may be considerable. Although there are several ways to reduce the time required, such as prospective data collection (Squillace *et al.* 1984) and computerized chart construction (Harding *et al.* 1989), the time factor is still an obstacle to their routine use. Can the expenditure of clinical time required to produce a life chart be justified? Specifically:

1. Does a life chart improve understanding of the patient?
2. Can a life chart influence management decisions?
3. Is a life chart an aid to treatment in its own right?

Understanding the patient

The problem of how best to summarize the patient's history, presents a major challenge to the psychiatrist. For clinical information to be efficiently interpreted and communicated it has to be condensed by one means or another. The commonly used means are the diagnosis, the formulation, and the problem list.

Diagnosis alone, even if multi-axial (as in DSM-III-R), offers an inadequate summary of an individual patient's plight (Birley 1990). Even when expanded by means of a formulation, which adds consideration of aetiology, or a problem list which enhances appreciation of the problems facing the patient, much potentially useful longitudinal information is lost.

The life chart is a valuable supplement to diagnosis, formulation, and problem listing. It adds the essential dimension of time (Vaillant 1987) and it reveals correlations between aetiological factors and treatment interventions from biological, psychological, and social realms (Engel 1980).

Natural history   The life chart provides a useful means of summarizing chronic or recurrent illness. It conveys the cumulative burden the patient has borne, and expands the narrow perspective or 'snapshot' impression of the single mental state examination or diagnosis. By depicting the longitudinal course of an illness the life chart reveals both the pattern and frequency of recurrence.

Cutler and Post (1982) examined hospital records of six patients with manic-depressive disorder from the pre-pharmacological era. By this means they were able to show great variation between individuals in the course of the disorder.

A study of schizophrenia by Harding *et al.* (1989) also demonstrated great variability between individuals with this diagnosis. A total of 296 patients from Vermont State hospital were followed for an average of 32 years after first admission. The authors used prospectively collected information to produce individual life trajectories and were able to distinguish differences in life course. Thus, within a single diagnostic group of patients there may be great variability in course, and a life chart can add considerably to the information conveyed by diagnosis.

Aetiology   The life chart can increase understanding of the aetiology of psychiatric disorder by illustrating the temporal relationship between onset of symptoms and possible causal factors. Figure 5 depicts the association between an increased frequency of binging in a woman with bulimia nervosa, and both academic failures and the ending of relationships. In their studies of patients with manic-depressive disorder Squillace *et al.* (1984) noted how detailed life chart analysis of their patients' illnesses often uncovered potential psychosocial precipitants of episodes of mood disorder, that would have otherwise been considered to be 'endogenous'.

The recurrent co-occurrence of symptoms with certain factors in a person's life, as in the cases above, may suggest an aetiological connection between the two, and a lack of association may suggest that a factor does

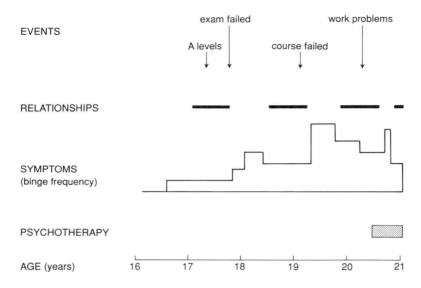

Fig. 5. A life chart showing the relationship of life events and the ending of relationships, with frequencies of binges in a woman with bulimia nervosa.

not play an aetiological role. However, caution must be exerted in interpreting such associations as correlation does not necessarily imply cause. Neither does a lack of coincidence firmly exclude a causal relationship as aetiological effects can be delayed.

Treatment response Life charts are useful in evaluating a patient's response to treatment, and particularly valuable in assessing the effect of prophylactic treatment on relapse rate. Schou (1973) used life charts to examine the effect of lithium carbonate on the course of manic depressive disorder. This method may be applied to other disorders. Figure 6 shows the relationship between relapse and the taking of antipsychotic medication in a woman with schizophrenia. Charts have also been used to examine the effect of psychosurgery for resistant bipolar affective disorder on relapse rate, frequency of hospital admission, and consumption of psychotropic medication (Poynton *et al.* 1988).

Prognostication As the longitudinal studies described above have demonstrated, persons with the same diagnosis may demonstrate great variation in life course. Since previous pattern of illness can predict future course (Squillace *et al.* 1984) a life chart can improve clinical prognostication above that based only on diagnosis. Furthermore, understanding the effect

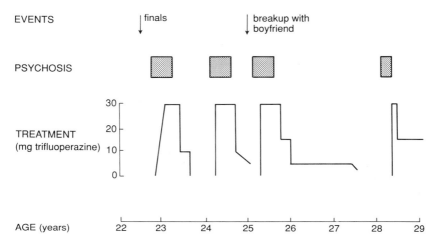

Fig. 6. A life chart showing the relationship between compliance with antipsychotic medication (trifluoperazine) and relapse in a woman with schizophrenia.

of specific psychosocial variables and treatment methods on the course of the illness will enhance any estimate so derived.

Clinical management decisions

The use of charts to enhance understanding of the patient is clearly valuable, but does this additional understanding alter management? In the case of manic-depressive disorder, life charts produced from retrospectively collected information have been shown to be superior to casual clinical estimates in judging the number of previous episodes of illness; to reveal previously unsuspected psychosocial precipitants of episodes of illness; to be useful in evaluating response to medication; and to predict subsequent course (Squillace *et al.* 1984). Hence, their use should allow management decisions to be 'fine-tuned' to the needs of individual patients. As a consequence, the choice of treatment may be altered, and more accurate advice given to patients and their families about the risks of particular life stresses. The author knows of no studies that have systematically evaluated the influence of life charts on management decisions, but clinical experience has demonstrated that such decisions are often modified after careful examination of a life chart.

Therapeutic tool

If the chart can be useful to increase the understanding, and change the behaviour of the clinician, can it also increase the patient's understanding of his or her own illness and thus change behaviour? By enlisting patients' co-operation in the creation of charts they become more aware of connections between life events and symptoms. Charts have been used in this way as an adjunct to psychotherapy. The chart shown in Fig. 5 was produced collaboratively by a patient with bulimia nervosa and her therapist. Psychotherapy subsequently focused on the association between certain loss events and an exacerbation of her bulimia, and resulted in a substantial improvement.

Charts may be used in a similar way to improve patient compliance with treatment. There are many reasons for non-compliance, but the author has found that discussion with the patient of a life chart that clearly indicates the relationship between non-compliance and relapse (e.g., Fig. 6) can help the patient appreciate the benefits of long-term medication.

Practical guidelines

The following is an outline guide to the construction and use of life charts.

Construction of charts

(1) The first stage in the construction of any life chart is the collection of accurate and relevant clinical data. The raw material of the life chart is clinical information of the sort collected in the standard psychiatric history. However, in collecting such information with the intention of producing a life chart, it is necessary to pay particular attention to the relative timing of symptoms, life events, and treatments received. One way to locate biographical events remote in time is to use anniversaries and other significant events, such as marriages and births, as temporal anchor points (Leighton and Leighton 1949).

Although existing clinical notes may provide an adequate basis for the construction of a chart, its quality can be enhanced if this, often incomplete clinical record, is supplemented by re-interviewing the patient, and an informant. Where maximum reliability is required, only information reported independently by two separate sources should be used in constructing the chart (Roy-Byrne *et al.* 1985).

(2) Before drawing a graphical chart it is advantageous to organize the collected information. To this end it may be split into broad categories; any number may be employed but three are commonly used: (a) the course of the illness, including symptoms and incapacity; (b) psychological, social, and biological 'stress factors' considered potentially relevant to its course; and (c) treatments, both physical and psychotherapeutic. This information may then be recorded in three columns, one corresponding to each of the above categories, the time scale running vertically.

(3) Once the clinician is satisfied that adequate data has been obtained, and the timing of items has been fixed as accurately as possible, the entries in the columns may be transferred on to a graph, the abscissa representing an appropriate time period. Although some charts may attempt to summarize an entire life history, it is often more useful to focus on a specific time period considered most relevant to the current problem. In order to ensure that the resulting chart is clear and easily understood it is necessary to be selective in transferring information from columns to chart, and also to consider how it should best be represented. For example, symptom severity can be portrayed as a continuous line that varies above and below a euthymic baseline. Various other techniques may be used to represent symptoms, treatments, and events; these include lines, columns, arrows, and symbols. The use of colour may further enhance clarity and ease of interpretation.

Use of charts

In addition to the uses described above a chart can provide a focus for a multidisciplinary team meeting. If drawn on an acetate sheet and projected it serves to keep discussion about the patient grounded in their life history. The drawing of charts is also a useful teaching exercise for students. Finally, the use of charts serves to highlight deficiencies in clinical records, and hence acts as a stimulus to their improvement; indeed a prospectively recorded life chart may be a valuable supplement to conventional notes.

Conclusions

Life charts were introduced at the beginning of the century. In their original forms they may now be thought to be merely historical curiosities. Recently, however, several groups of researchers have developed a form more suited to the needs of the modern clinician. These charts can portray the natural history and multifactorial aetiology of chronic and recurrent psychiatric conditions, and hence improve the recording and communication of clinical information. They may also be used as a tool in psychotherapy, and in making optimum use of prophylactic treatments. The potential of life charts in clinical practice and research, both in psychiatry and other medical specialties, has not yet been fully exploited.

References

Birley, J. L. T. (1990). DSM-111: From left to right or from right to left? *British Journal of Psychiatry*, **157**, 116–18.

Cutler, N. R. and Post, R. M. (1982). Life course of illness in untreated manic depressive patients. *Comprehensive Psychiatry*, **23**, 101–15.

Engel, G. L. (1980). The clinical application of the biopsychosocial model. *American Journal of Psychiatry*, **137**, 535–44.

Harding, C. M., McCormick, R. J., Strauss, J. S., Ashikaga, T., and Brooks, G. W. (1989). Computerized life chart methods to map domains of function and illustrate patterns of interactions in the long-term course trajectories of patients who once met the criteria for DSM-III schizophrenia. *British Journal of Psychiatry*, **155**(Suppl. 5), 100–6.

Kraepelin, E. (1913). *Psychiatrie*. Johan Ambrosius Barth, Leipzig.

Leighton, A. H. and Leighton, D. C. (1949). *Gregorio, the hand-trembler: A psychobiological personality study of a Navaho Indian*. Peabody Museum, Cambridge, Mass.

Meyer, A. (1917/1952). Mental and moral health in a constructive school programme (Reprinted 1952). In *The collected papers of Adolf Meyer* (ed. E. E. Winters), pp. 350–70. Johns Hopkins Press, Baltimore.

Post, R. M., Roy-Byrne, P. P., and Uhde, T. W. (1988). Graphic representation of the life course of illness in patients with affective disorder. *American Journal of Psychiatry*, **145**, 844–8.

Poynton, A., Bridges, P. K., and Bartlett, J. R. (1988). Resistant bipolar affective disorder treated by stereotactic subcaudate tractotomy. *British Journal of Psychiatry*, **152**, 354–8.

Roy-Byrne, P., Post, R., Uhde, T., Porcu, T., and Davis, D. (1985). The longitudinal course of recurrent affective illness: life chart data from research patients at the NIMH. *Act Psychiatrica Scandanavica*, **71**(Suppl. 317), 5–34.

Schou, M. (1973). Prophylactic lithium maintenance treatment in recurrent endogenous affective disorders. In *Lithium: its role in psychiatric research and treatment* (ed. S. Gershon and B. Shopsin), pp. 269–94. Plenum, New York.

Sharpe, M. (1990). The use of graphical life charts in psychiatry. *British Journal of Hospital Medicine*, **44**, 44–7.

Squillace, K., Post, R., Savard, R., and Erwin-Gorman, M. (1984). In *Neurobiology of mood disorders* (ed. R. Post and J. Ballinger), pp. 38–59. Williams & Wilkins, Baltimore.

Trethowan, W. H. (1977). Music and mental disorder. In *Music and the brain* (ed. M. Critchley and R. A. Henson), pp. 398–432. Heinemann, London.

Vaillant, G. E. (1966). Twelve-year follow-up of New York narcotic addicts. *New England Journal of Medicine*, **275**, 1282–8.

Vaillant, G. E. (1987). Time: An important dimension of psychiatric epidemiology. In *Psychiatric epidemiology progress and prospects* (ed. B. Cooper), pp. 167–77. Croom Helm, London.

19

Research questionnaires in routine clinical practice: are they useful?

CHRIS THOMPSON

Introduction

The dilemma contained in the title of this chapter is inherent in the different aims and objectives and therefore the different methods of clinical psychiatry and research. In research, the objective is to demonstrate a result in highly standardized terms to others who were not engaged in the research. In clinical practice, the aim is to help people to get well. As their own report is an essential element in judging when this has been achieved the use of standardized questionnaires is not usually necessary or helpful. The judgement is between the clinician and the patient. However, the point of this chapter is to explore those clinical situations in which there might be some utility in using one of the standardized questionnaires which have been developed for research purposes.

Categories of questionnaire

When we refer to questionnaires in this respect we are including classification schemes with their own interview schedules and observer-completed ratings of severity of an illness, as well as the more common use of the term to refer to a self-rating inventory. A generic term for all of these standardized procedures might be 'research instruments'.

A huge number of research instruments is now available for making operational definitions of psychiatric disorders of various kinds and for making judgements about severity (Thompson 1989). There continues to be a great deal of activity in producing yet more. Indeed, a new journal, the *International Journal of Methods in Psychiatric Research* has been set up to carry some of this work, which often finds no outlet elsewhere because it is considered to be too technical for the general psychiatry journals. As Hall (1980) has observed: 'in practice it is extremely rare to find references to articles on scale construction other than the first published article'.

This plethora of instruments and the relative lack of documentation about each can cause some confusion. There are for example at least six versions of the Hamilton Rating Scale for Depression, and altogether Thompson (1989) found seven more scales with an observer-rated component for the severity of depression and a further seven which were entirely self-rated. In addition, there were six scales which measured depression as one component of a mixed mood disorder.

When we turn from severity to diagnosis, the two great classification schemes, International Classification of Diseases (ICD) and the Diagnostic and Statistical Manual for Mental Disorders (DSM) are the most widely used clinically. However, in research, a range of other instruments are sometimes used, some of which are fairly broad (e.g., Research Diagnostic Criteria), whereas others are relevant only to small differential diagnostic questions, such as the Newcastle Index for endogenous depression.

The properties of rating scales and their relevance to clinical practice

There are three essential properties of any standardized instrument which have to be adequate before it can be used.

1. *Validity*

Does the instrument measure what it is supposed to be measuring. There are various ways of estimating this. Do the items in the instrument look relevant to the syndrome as you know it clinically? This is 'face validity'. Do the scores correlate well with another widely used instrument which is generally accepted to be valid? This is 'concurrent validity'.

2. *Reliability*

If the instrument is used repeatedly or by different raters does it each time give much the same result? The former test of reliability is called 'test-retest reliability' while the latter is 'inter-rater reliability', which has been given great prominence in the psychiatric literature because of the well-documented unreliability of unaided clinical diagnostic practice.

These first two properties are essential elements of any measuring instrument and are the most important in research terms. However, the third (below) is equally important when it comes to clinical practice.

3. *Utility*

This is the likelihood that the result of the test will affect your clinical judgement. To be useful, a test has to have a high positive predictive value, for example, to indicate that the individual is really psychiatrically ill in the case of screening test, or that he or she will really respond to antidepressants in the case of a predictive test. In each case, this involves finding the best of several scores to act as a threshold in order to dichotomize the information.

In research, one works with groups but in clinical practice with the individual. So the utility of a clinical test is its ability to make accurate predictions and thus affect a clinical decision about diagnosis or treatment. The dexamethasone suppression test is a good example of a test which appears in research terms to be reliable and valid but to have low utility because it was not found to alter clinical decision-making.

This, then, is an actuarial meaning of the term utility (Rowland 1985). A test's usefulness will also depend upon its ease of use in a busy clinic. In this respect, self-rating scales are more useful than clinician-rating scales but unfortunately they lack more in the area of reliability. Another point to think about is the duration of training required to use an instrument like the Present State Examination (about one week even for a trained psychiatrist). The more reliable instruments tend to be less clinically useful because they take more training and take longer to complete on each patient.

Why standardize assessments?

For the reasons outlined above, it has not been the accepted practice to use rating scales regularly to guide diagnosis or treatment in routine clinical practice. Indeed, this has been the case across all of clinical medicine, although physicians and others often have laboratory markers which act as proxies for the clinical condition itself. Some clinicians may have guidelines in their head when making diagnoses which tend more towards the DSM-III-R or the ICD 9 concept of a disorder. But apart from this 'rule of thumb' use of operational criteria few would routinely use say the 'Structured Clinical Interview for DSM-III-R' or even have open on the desk the glossary for ICD 9. Indeed, it is almost considered unprofessional practice to do so, because this may not convey to patients a sense of confidence in the abilities of their psychiatrist! Psychiatric training inculcates flexible interviewing, and quite rightly so. Effectively, the only reason to standardize an assessment is when it is necessary to communicate reliably with others. The others might be the research community, the multidisciplinary team, hospital managers, or even the patient who is lacking in insight. In the future, it may be necessary to be able to do this for clinical practice because of the advent of frequent routine audit. A service which cannot demonstrate efficacy might not survive.

Some examples of when standardized measures might be helpful

The examples below have been divided up into broad categories but each category overlaps with the others because the need to communicate is the main reason for standardization.

1. *Assessment of change in chronic conditions*

Where the time course of change is over a long period, frequent retro-spectives may be necessary to judge the effectiveness of treatments. Any standardized rating scale appropriate to the condition could be used in this context. Commonly used and fairly simple scales include the Brief Psy-chiatric Rating Scale in Schizophrenia (Overall and Gorham 1962), The Hamilton Rating Scale for Depression (Hamilton 1960), the Clinical Anxiety Scale (Snaith *et al*. 1982), the Hospital Anxiety and Depression Scale (HAD: Zigmund and Snaith 1983) for mixed states and the Beck Depression Inventory (Beck *et al*. 1961).

An advantage of having regular ratings of the severity of the condition is that they can be discussed with patients if they have any concern about the treatment in order to enhance adherence to treatment.

Behavioural therapists are encouraged routinely to use linear scales in assessing the severity of the target complaint, such as a phobia or ritual (Gelder and Marks 1966). There is a good correlation between the therapist's and the patient's assessments (0.74–0.81 for main phobia). The develop-ment of behavioural therapy training programmes from research in behavi-oural intervention has undoubtedly been responsible for establishing this routine use of measurement in behavioural therapy.

2. *Comparisons between patients*

Another use is where assessments are complex and comparisons have to be made between patients; for example, in the assessment to determine which patients in a long-stay ward are most suited to transfer to a hostel. Here, the results can also be used to carry information from the ward team to the hostel team.

The most commonly used scale for this is the 'Rehab' or 'Hall and Baker' scale. This consists of seven questions about difficult or embarrassing behaviour, such as: 'Did the patient shout or swear at others?' The time period of enquiry is the last week. In addition there are visual analogue scales for behaviour which can be rated on a continuum from normal to abnormal, such as: 'How clearly did the patient speak?'

The manual for the scales (Baker and Hall 1983) indicates a number of uses. These include measurement of behaviour before, during, and after treatment (e.g., in token economies), as well as the identification of those who might have potential for living outside hospital, or grading patients with similar levels of self-care for appropriate placement within the hospital. The large variety of clinical situations which can be encountered is acknowledged in a footnote suggesting that the words 'patient' and 'hospital' can be modified to 'client' and 'hostel'. The effect of this difference of environment on the validity of the instrument would be critical in assessing its use in a research project, but in clinical practice, a more pragmatic view can be taken. In other words, is it found to be useful by the staff in making decisions?

Untrained observers can use the scales because they concentrate not on psychopathology but on observed behaviour. A glossary is available for standardization of definitions.

A related situation often occurs where the patient does not talk, and information about mental state is obtained from several sources or from an informant, such as in mental handicap. The MRC Handicaps and Behaviours Schedule (Wing and Gould 1978) was devised to assess, with an informant, the clinical state of mentally handicapped patients. It deals with developmental skills and abnormal behaviour. Inter-rater reliability is reported to be high (70 per cent). A brief version is available for clinical practice. It is recommended for use in service planning exercises.

3. *To overcome diagnostic uncertainty*

Scales may be used to confirm a diagnosis where there is doubt among the team members. This situation may arise not infrequently where there is disagreement about the correct diagnosis. Beck *et al.* (1962) showed the unreliability of unaided clinical diagnosis 30 years ago and it has been repeated many times since. The resort to a standardized instrument may not resolve the disagreement but it provides the debate with an external arbiter (i.e., the scale) and this is backed up by the information available about the validity of the scale against other scales and against groups of clinicians in the centre where the scale originated. Examples of instruments that could be used would be the DSM-III-R interviews (APA 1987) or the Present State Examination (Wing *et al.* 1974). The main problem with this approach is that the classification schemes which have been used for research, or developed from them, are designed to make highly reliable diagnoses. In clinical practice, this leads to the exclusion of large numbers of doubtful cases. In the end, the ragbag of unspecified illnesses is sometimes as large as that containing classified conditions. This is not useful and a less tightly controlled system of classification is usually more helpful. This general point is important throughout our consideration of the transfer of research 'technology' to clinical practice. The patients most likely to need their diagnosis checking against a standardized instrument are just those most likely to fall into the unspecified group.

4. *Screening for psychiatric disorder in non-psychiatric settings*

An extensive literature on the screening of general practice patients has grown up around the work of Shepherd *et al.* (1966) and Goldberg and there are many instruments which have been used. Pre-eminent among these is the General Health Questionnaire (GHQ: Goldberg 1972, 1978), which has more or less superceded the older instruments, such as the Cornell Medical Index. The GHQ performs well when the validation is the Clinical Interview Schedule carried out by a trained rater. It is easy to complete and can therefore be used with some reliability to identify for closer inspection those patients in a general practitioner's or out-patient clinic who would repay closer attention to their psychiatric state. Why then is it not more widely used in practice? The main reason is that many general practitioners (GPs) and physicians would not know what to do with the patients ascertained as having a psychiatric component to their illness and many think they would be very time-consuming to deal with. Recent studies showing the relationship between depressive symptom severity and response to antidepressants in general practice do suggest that more screening might have practical advantages over and above the purely epidemiological academic ones (Hollyman *et al.* 1988). Similarly, case identification for counselling can be helpful. When case identification for intervention is required it is often appropriate to raise the threshold score so that the positive predictive value is enhanced. This minimizes false positives at the expense of missing some true positives (Goldberg 1986).

However, routine screening for mental disorder is not yet a part of the GP's contract (cf. cervical screening), reflecting continuing scepticism about the effectiveness or cost effectiveness of the exercise.

Wilkinson and Barczak (1988) used both the GHQ and the HAD in general practice and found that the positive predictive value of the HAD was 81 per cent and that of the GHQ was 77 per cent. The HAD was also more sensitive than the GHQ (90 per cent vs. 77 per cent). An additional complication occurs where there are pronounced symptoms of physical illness. Here, it may be necessary to alter the threshold for detection in order to maintain a reasonable positive predictive value (Goldberg 1985). For routine screening then, the HAD might be more appropriate than the GHQ, but the authors also make the point that, for a variety of operational reasons, the GP's diagnoses might be more appropriate in many instances where they differ from the HAD or the GHQ. General practitioners do, however, miss a large fraction of the cases screened as positive by the HAD or GHQ, and so the best method in a screening clinic would be for the GP to make a final diagnosis, taking the questionnaire results into account.

5.   *The monitoring of clinical progress*

Questionnaires can be used like a temperature chart to follow improvements in acute illnesses. This differs slightly from their use in following chronic illness where the scores act as a reminder of what the patient was really like several months ago so that progress can be seen and confirmed, or otherwise. In the acute situation, the scores can be used to demonstrate improvements brought about apparently by one treatment rather than another. This can then be used to communicate within the clinical team. The ratings are usually made by nurses.

An example of this is the Bunney Hamburg Rating Scale for longitudinal observation of a wide variety of behaviour associated with abnormal mental states (Bunney and Hamburg 1963). However, it has been little used outside its place of origin in Bethesda. The authors proposed that it be completed every eight hours to get a picture of the behaviour. They found that nurses were able to make the ratings, but with a variable degree of reliability from 0.11 to 0.83 for the 24 items of the scale, each given a 15 point rating. Although this scale has not been much used in research it might find a place in the regular rating of behaviour in in-patient settings where acute change is expected. In view of its potential usefulness it is surprising that there is no literature on modifications and improvements which might make it more acceptable to other units.

An important point is not to assume that nurses can use rating scales designed for psychiatrists. Although nurses will often be reliable when judged against other nurses using, say, the Hamilton Depression Scale, they give very unreliable ratings when judged against doctors. This is a function of the nurse training relying more on the assessment of overt behaviour, and their lack of training in the methods of descriptive psychopathology.

Whereas nurse ratings are a common way to acquire data, self-ratings are also cost effective in terms of time. Snaith has proposed that the Hospital Anxiety and Depression Scale (HAD) could be used like a temperature chart to plot the recovery from depression (Zigmund and Snaith, 1983; Snaith 1985). An example of this is given in Fig. 7. The advantage of the HAD is the relative absence of somatic symptoms in the item list, which, if present, could give a falsely positive impression of recovery because of the masking of symptoms by the side-effects of the older tricyclic antidepressants. Only one item, 'feeling slowed down', is responsive to somatic complaints.

The problem with using self-rating scales is that their validity is low because the criteria on which ratings are made are specific to the individual patient who has only his or her own experience to go on. Thus, maximum severity is interpreted differently in each case. Nevertheless, this does not invalidate their use to follow individual patients over time and the charting of HAD scores on a 'temperature chart' is one way to achieve this. On the same form can be included the investigations and treatments so that any improvement can be seen in relation to these other events.

6. *Communication between staff*

Research instruments can be used to communicate regular events on a ward from nurses to others about, for example, violent incidents. Two such scales have been reported which are essentially standardizations of the violent

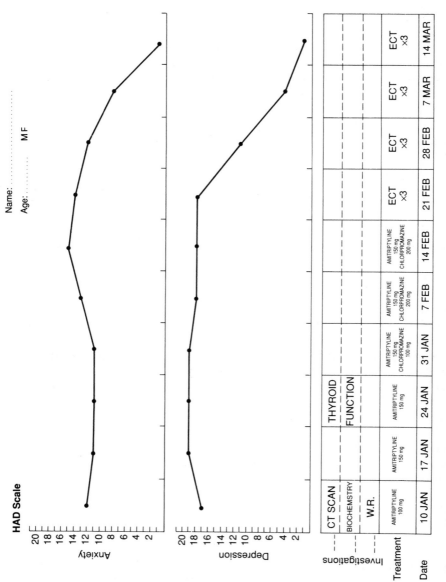

Fig. 7. Chart showing a patient's progress according to Hospital Anxiety and Depression Scale scores during treatment for depression.

incident report sheet in use in many hospitals. The Overt Aggression Scale (Yudofsky *et al.* 1986) is a checklist containing ratings of four categories of violence: against others, objects, the self, and verbal aggression. The timing of each incident and the intervention are given. The Staff Observation Aggression Scale (Palmstierna and Wistedt 1987) is very similar. Both have good inter-rater reliabilities and can be used to chart the course of improvement in individual cases, but more importantly to demonstrate the levels of disturbance on psychiatric wards, levels which are often only obvious to the nursing staff.

7. *Treatment prediction*

In a few cases where treatment decisions are not obvious it may be helpful to refer to scales specifically designed from research data to distinguish likely responders from non-responders. An example of this is the ECT prediction scale of the Newcastle group (Carney *et al.* 1965). Individual items are rated positively or negatively for likely response. A total score on the 10 items of 1 or more indicates response and of 0 or negative values indicates no response. The use of such scales might improve the response rates to ECT in a unit if used regularly, but in clinical practice many other constraints apply to treatment. The utility of the scales in this context has not been tested.

8. *Clinical training*

Some of the best scales can be successfully used in clinical training. The Present State Examination is a good example of this as its glossary is extensive and the interview technique is not so highly structured that it leaves no room for initiative. One of the purposes is explicitly stated to be clinical training (Wing *et al.* 1974).

Clinical guidelines

When clinicians are considering transferring the measurement techniques of psychiatric research to the clinical situation they have to be aware of a number of pitfalls, especially the training requirements for achieving reliability and validity and the reduction in the utility of the instruments when they are required to make predictions about single cases. However, there are occasions when they can be useful and some of these have been described. In particular they can help in the following situations.

1. Communication between team members, especially when they work in the community and may not meet each other frequently. In that situation a regularly administered questionnaire filed in the notes will reflect the progress of the patient for all team members to see.

2. The routine use of minimal standard recordings of outcome might become necessary in a more competitive health care 'market'. Then it will be important for clinicians to ensure that they are using outcome measures which adequately reflect clinical opinion. There is currently a consensus that the techniques for such detailed outcome measures are not sufficiently advanced to be able to guide resource management.

3. The results of self-ratings can be used in the interview with the patient. This has a number of effects. The item content of the questionnaire can reassure the patient that his or her condition is not unique. The item content also guides the clinician to questioning in some of the most important areas of the condition if the questionnaire has been well chosen. The results can be used as a partial guide to treatment.

   There are problems in the use of information in this way. Some patients are extreme raters and little of use can be gleaned from their self-rating scales. Scores on any kind of rating scale should not be used as a diagnosis.

4. Questionnaires and interviews can be very useful in training provided that the trainee is already sufficiently advanced and has good enough supervision to make proper use of the experience.

5. There are also several specialized settings in which research instruments can be used in clinical practice and these are covered in the body of the chapter. They include service planning and screening.

References

APS (American Psychiatric Association) (1987). *Diagnostic and Statistical Manual for Mental Disorders* (3rd edn, revised). American Psychiatric Association, Washington DC.

Baker, R. D. and Hall, J. N. (1983). *REHAB: a multipurpose assessment instrument for long-stay psychiatric patients.* Vine Publishing, Aberdeen.

Beck, A. T., Ward, C. H., Mendelson, M., Mock, J., and Erbaugh, J. (1961). An inventory for measuring depression. *Archives of General Psychiatry*, **4**, 561–71.

Beck, A. T., Ward, C., Mendelson, M., Mock, J., and Erbaugh, J. (1962). Reliability of psychiatric diagnoses: a study of consistency of clinical judgements and ratings. *American Journal of Psychiatry*, **119**, 351–6.

Bunney, W. E. and Hamburg, D. A. (1963). Methods for reliable longitudinal observation of behavior. *Archives of General Psychiatry*, **9**, 280–94.

Carney, M. P., Roth, M., and Garside, R. F. (1965). The diagnosis of depressive syndromes and the prediction of ECT response. *British Journal of Psychiatry*, **111**, 659–74.

Gelder, M. G. and Marks, I. M. (1966). Severe agoraphobia: a controlled trial of behaviour therapy. *British Journal of Psychiatry*, **112**, 309–19.

Goldberg, D. P. (1972). *The detection of psychiatric illness by questionnaire* (Maudsley Monograph 21). Oxford University Press.

Goldberg, D. P. (1978). *Manual of the General Health Questionnaire.* NFER Nelson, Slough.

Goldberg, D. (1985). Identifying psychiatric illness among medical patients. *British Medical Journal*, **291**, 161–2.

Goldberg, D. (1986). Use of the General Health Questionnaire in clinical work. *British Medical Journal*, **293**, 1188–9.

Hall, J. N. (1980). Ward rating scales for long stay patients: a review. *Psychological Medicine*, **10**, 277–88.

Hamilton, M. (1960). A rating scale for depression. *Journal of Neurology, Neurosurgery and Psychiatry*, **23**, 56–62.

Hollyman, J. A., Freeling, P., Paykel, E. S., Bhat, A., and Sedgwick, P. (1988). Antidepressant treatment in general practice: a placebo controlled trial of amitriptyline. *Journal of the Royal College of General Practitioners*, **38**, 393–7.

Overall, J. E. and Gorham, D. R. (1962). The Brief Psychiatric Rating Scale. *Psychological Reports*, **10**, 799–812.

Palmstierna, T. and Wistedt, B. (1987). Staff Observation Aggression Scale, SOAS: Presentation and evaluation. *Acta Psychiatrica Scandinavica*, **76**, 657–63.

Rowland, L. (1985). Assessment. In *Psychological applications in psychiatry* (ed. B. Bradley and C. Thompson), pp. 223–45. Wiley, Chichester.

Shepherd, M., Cooper, B., Brown, A. C., and Kalton, G. W. (1966). *Psychiatric illness in general practice.* Oxford University Press.

Snaith, R. P. (1985). A mood chart for use in clinical practice. *British Journal of Social and Clinical Psychiatry*, **3**, 16–18.

Snaith, R. P., Baugh, S. J., Clayden, A. D., Hussain, A., and Sipple, M. A. (1982). The Clinical Anxiety Scale: an instrument derived from the Hamilton Anxiety Scale. *British Journal of Psychiatry*, **141**, 518–23.

Thompson, C. (1989). *The instruments of psychiatric research.* Wiley, Chichester.

Wilkinson, M. J. B. and Barczak, P. (1988). Psychiatric screening in general practice: comparison of the general health questionnaire and the hospital anxiety and depression scale. *Journal of the Royal College of General Practitioners*, **38**, 311–13.

Wing, J. K., Cooper, J. E., and Sartorius, N. (1974). *The measurement and classification of psychiatric symptoms.* Cambridge University Press.

Wing, L. and Gould, J. (1978). Systematic recording of behaviours and skills of retarded and psychotic children. *Journal of Autism and Childhood Schizophrenia*, **8**, 79–97.

Yudofsky, S. C., Silver, J. M., Jackson, W., Endicott, J., and Williams, D. (1986). The Overt Aggression Scale for the objective rating of verbal and physical aggression. *American Journal of Psychiatry*, **43**, 35–9.

Zigmond, A. S. and Snaith, R. P. (19837. The Hospital Anxiety and Depression Scale. *Acta Psychiatrica Scandinavica*, **67**, 361–70.

20

Doctors with emotional problems: how can they be helped?

JONATHAN CHICK

Introduction

Mortality statistics in many countries show three conditions from which doctors tend to die more frequently than the general population: *suicide*, *liver cirrhosis* (much of which in the Western world is alcohol-related), and *accidents*. The first two, and perhaps the accidents also given the vagaries in the certification of suicides, point to psychiatric illness.

Studies of the *suicide rate in physicians* are not completely unanimous in finding an excess, and methodological problems in interpreting the occupational mortality data are described by several authors (e.g., Arnetz *et al.* 1987; Roy 1985). In the UK, however, the standardized mortality rate (SMR) for suicide in doctors has remained high throughout the twentieth century, except for the 1930s, when the national suicide rate increased dramatically during the economic depression. It is currently twice the rate for other professionals. In Sweden, the suicide rate in male physicians is not raised in comparison with the general population but is raised in comparison to others in the population with higher education (Arnetz *et al.* 1987).

Women physicians are at even greater risk of suicide than are men. Pitts *et al.* (1979) studied the information about cause of death in American physicians during the period from 1967 to 1972, and found that the suicide rate in the 751 female physicians who had died was 6.52 per cent, four times higher than the rate for white women over the age of 25. The recent 10 year prospective Swedish study found an SMR for female physicians of 5.7 (Arnetz *et al.* 1987).

Murray (1977) found that overall *psychiatric hospitalization rates* for Scottish male doctors were more than twice those of Social Class I controls. This difference was accounted for by the excess in the diagnoses of drug dependence, alcoholism, and depression. Studies of hospitalization in Canada, the US, and England and Wales have found the same pattern.

The *cirrhosis mortality* SMR in doctors has fallen in the UK recently, from 350 in 1962, to 311 in 1970–2, and yet further, to 115, in 1982–3 (OPCS 1986). The possibility that hepatitis B exposure might have accounted for the previous high rates is not supported by the lower than average cirrhosis rate in nurses, and thus alcoholism has usually been deemed to be the cause. No studies of trends in physicians' alcoholism morbidity rates are available to indicate whether perhaps by drinking less alcohol in recent years British doctors have lowered their incidence of this condition. An explanation for the fall in their cirrhosis death rate could conceivably be that alcoholic doctors now have a better outcome than two decades ago, and this will be discussed below.

Differences between specialties

The weight of the evidence in North American studies is that the suicide rate is higher among psychiatrists than among doctors who choose other areas of work. However, the well-conducted Swedish study already mentioned found that out of the 30 physician suicides over a 10 year period, surgeons had the highest relative risk (Arnetz *et al.* 1987).

Alcoholism, however, does not seem to be the prerogative of the psychiatrist. Murray (1976) found that only 6 per cent of the psychiatrists referred to the Maudsley Hospital in London were diagnosed as alcoholic, compared to 33 per cent of surgeons and 37 per cent of general practitioners. Extraversion and gregariousness are predictors of alcoholism when college samples are followed up, and perhaps psychiatry tends to recruit less of these personality types than surgery or general practice.

The causes of psychiatric illness in doctors

There is data to support both 'selective recruitment' (medicine attracts some vulnerable individuals) and an 'exposure' factor (stress, or other aspects of the work or of a doctor's life) to explain increased psychiatric illness in doctors.

The Harvard sophomore study of 268 college students was able to follow up 46 men who later went on to medical school, and chose 79 controls (fellow students) who did not (Vaillant *et al.* 1972). In the repeated interviews in the next three decades, doctors, especially those involved in direct patient care, were more likely than controls to report being unhappily married, to abuse drugs, and to have psychiatric treatment. These problems were associated with having had an unsatisfactory childhood, and greater instability at college, rather than any particular stress related to medical practice. It was concluded that there is a vulnerable minority among medical students, who may enter the profession as an attempt to be involved in caring relationships to compensate for their own unsupported childhoods and instability: 'It is in the physician with a barren childhood who is feeling overly burdened by dependent patients that trouble arises' (Vaillant *et al.* 1972).

Waring (1974) similarly concluded that family history, life experience, and personality before entering medical school, accounted for much of the psychopathology seen in doctors. A report in 1869 of the future careers of 1000 graduates of St. Bartholomew's Hospital Medical School, London, led James Paget to note that, of the 56 pupils who had 'failed entirely', 10 had done so 'through the continuance of the same habits of intemperance or dissipation, as had made us, even while they were students, anticipate their failure' (Collier and Beales 1989). Today male medical students do not report more alcohol consumption than other young men, though women medical students do drink more than their female peers (Collier and Beales 1989).

It seems to be only among women medical students that a higher than expected suicide rate is found. Surveys of medical students find that, compared to employed people of the same age in the general population, symptoms of anxiety and depression are more commonly reported by female and perhaps slightly less so by male students (reviewed in Surtees and Miller 1990). If a comparison is made with other undergraduate populations, such as law, pharmacy or dental students, it is seen that other disciplines also have a stressful rite of passage (Roy 1985). A higher rate of stress symptoms than in the general population also continues to be seen after graduation (Firth-Cozens 1989).

Part of the high doctors' suicide rate often found in the literature may be accounted for by their greater knowledge of lethal methods, as with pharmacists. Another undoubted contributary factor is alcoholism, follow-up studies of alcoholic physicians finding high rates of suicide (Roy 1985). For example, Murray (1976) followed 36 alcoholic doctors over five years and found that four had died by their own hand.

The American Medical Association (1987) gathered information from the colleagues, friends, and relatives of 142 doctors who died of suicide between 1982 and 1984, and compared this with information about 101 physicians matched for age and sex who had died of other causes. Most (94 per cent) were thought to have taken their lives because of mental pain, and some (26 per cent) also because of physical pain. There was a trend, not statistically significant, for the doctors who died of suicide to have had more difficult, emotionally draining patients in recent years. Alcohol and drug abuse were factors, as were previous psychiatric illnesses. Many (42 per cent) had been seeing a mental health specialist at the time of death, suggesting either an underestimating of the risk of suicide by therapists, or perhaps that there was an intractable chronic illness for which it was difficult to find a cure (58 per cent had been hospitalized for psychiatric treatment more than once). Physicians who had died of suicide were more likely to be thought of as critical of others, and highly critical of themselves. They tended to have fewer social supports. Alcohol abuse and depression were more common in their family histories.

The ready availability of drugs is an obvious explanation for the increase in drug dependence among doctors. There are a number of reasons put forward to explain why alcoholism is, or has been at least until recently, common in doctors. As well as usually being able to afford alcohol, some doctors work with little supervision, or in work settings in which there is no superior to comment on failing performance.

However, there are, of course, some aspects of work with patients which can lead to great anxiety, or despair, and which may lead either to use of drugs or alcohol as an escape, or to depression. Contact with death can disturb a doctor even after many years of practice. The fear of making mistakes, or having made a mistake, is something many doctors know. A patient's legal complaint, or a negligence investigation, can have a profound effect on a doctor's self-confidence, and his mood, as he or she goes through the phases of constant rumination on what happened, of anger, of fear of losing his professional authority or even losing his livelihood. Depression can arise when the work is unsatisfying because of the pressure of sheer numbers of patients to be seen, when feelings of being taken for granted by patients or the managing authority may fester and lead to demoralization or resentment.

Stress related to work has been studied in detail in contemporary British general practice by Branthwaite and Ross (1988). Questionnaires returned from 408 of 632 doctors revealed that 56 per cent experienced uncertainty or insecurity in their work, 38 per cent social isolation, 38 per cent poor relationships with other doctors, and 22 per cent disillusionment. These negative comments were more frequent in doctors who worked alone or in small groups, perhaps without support staff, and who had large practice lists.

Doctors' marriages may not be as supportive as they could be, if the diversion of energy and time away from the marriage and into helping patients has fostered resentment or loss of self-esteem in the spouse, with a deterioration of communication and loss of expressed affection. Doctors' wives sometimes feel profoundly neglected, a sad matter which also shows in the high suicide rate in wives of doctors compared to wives of other professionals reported by Sakinofsky (1980) in the UK.

Arnetz *et al.* (1987) found that women with higher education have an elevated suicide rate, approaching that of women doctors. This suggests that suicide among female physicians may be partly related to factors common to all women who aim for a career, such as the strain of combining the roles of wife, mother, and profession. The divorce rate is higher for women than men doctors (Lorber and Ecker 1983). Depression seems to be more common in female than male junior doctors and the contributants to this discrepancy have recently been discussed (Godlee 1990). A UK study of women junior doctors found that the most frequent stressors described were overwork and conflict between career and personal life. However, there was also a high correlation between emotional disorder, the reporting of a poor relationship with the consultant, and worry over making clinical decisions (Firth-Cozens 1990).

Asking for help

'There were many dismal stories about practitioners whose competence to practise had plainly been impaired by illness, often psychiatric, but who soldiered on to the distress of their families and to the alarm of their colleagues' (Rawnsley 1985). There is an expectation felt by doctors themselves that they should not need help, that they should be able to sort out their own problems—'Physician heal thyself'. This is probably especially true in the UK, where in many sections of society, seeking psychological help is seen as weakness. This perpetuates *denial* of the illness. Being on the receiving end of help or advice is something many doctors find difficult, but airing this feeling may help a doctor cross the first hurdle.

Medical knowledge is no barrier to a doctor developing disabling hypochondriasis, when palpitations and dizziness are in fact due to the panic attacks of the anxiety or depressive state he or she will not acknowledge. Severe depressive illness may involve profound loss of insight. However, with drug and alcohol problems there are other dimensions to the denial which prevents seeking help. There is the fear of moral stigmatization. There is also the reluctance to admit any course of action which may lead to having to make a decision to reduce or terminate a habit which has been or still is enjoyable or felt as a necessary aid to coping.

Family and colleagues of the drug- or alcohol-abusing doctor may be drawn into *covering up*. Colleagues may turn a blind eye, even though their resentment grows as extra work accumulates for them. Lateness, unreliability, the smell of alcohol at work, tremor at the operating table—these signs are dismissed as 'stress'. No one wants to confront him or her. The spouse perhaps feels guilty and is just as afraid of the stigma of acknowledging an alcohol problem as the drinker. Colleagues are easily put off by the drinker making them feel like a wet blanket, spoiling the fun, or like meddlesome interferers who should mind their own business. Or there is a misplaced belief that 'loyalty' must override other considerations. An eminent surgeon's alcoholism was described thus (Strega 1978):

. . . the extent of his drinking was acknowledged throughout the hospital. The subject was rarely discussed openly and then only with a mixture of jocularity ('Guess what he did next?') and hopelessness ('Well, how could I stop him?'). Nothing it seemed could threaten his inviolability, supported as he was by a pyramid of housemen, registrars, lecturers and research fellows. If one man had stepped out of line the whole fragile construction would have fallen; but whether out of loyalty to the gifted surgeon he once was or whether out of fear for the power of a referee, nobody moved a muscle.

It can take courage and compassion to act to help a sick colleague to get well.

Persuasion/coercion

Colleagues think that reason will eventually prevail. When it does not, and the miracle they pray for does not happen, and anger and resentment are allowed to build up, there is eventually a crisis leading to violent rejection perhaps with little understanding. Complaints are made. The doctor loses his reputation, perhaps his career; his alcoholism or depression gets much worse; suicide may seem a way out (Crawshaw *et al.* 1980).

At this point, when disciplinary action is considered, the employing authority, the General Medical Council (UK), the State Board (US) or other board of accreditation, may require the doctor to have a psychiatric or medical assessment and follow recommended treatment. However, if family and colleagues or managers had acted sooner, matters need not have gone so far.

Records should be kept of specific events which have led to the conclusion that the individual must get treatment. The timing for approaching a colleague who appears to be ignoring a psychiatric illness or an addiction sometimes needs to be planned. Occasionally, an approach will only be successful when there has been a further incident. The approach should be at a time when he or she is sober. Colleagues, administrator, and family, and perhaps other friends might be brought together, so that their voice, expressing caring as well as facing reality, is united. Under this co-ordinated pressure, the sick doctor may find it harder to refuse to take the next step, which might be, for example, to see his or her own general practitioner, to see an adviser recommended by the national or regional specialty colleagues, or to go for a psychiatric assessment. One should not expect or accept promises: these will be futile, especially with addictions.

Establishing at this point an agreement about openness between the parties concerned can later be invaluable: the doctor may otherwise suffer in the way 'special patients' suffer, because of the smoke-screen of excessive secrecy—dead of night phone calls giving information, 'which on no account is to be passed on'. The colleagues or employer should state that they will pass the information they have to the assessing psychiatrist. This reduces the chance that the individual deceives the specialist into passing him or her as fit and well.

It may be easier for unwell doctors to agree to assessment, or to seek treatment, outside their own district. Doctors are sceptical, and competitive, by nature and by training. They may not welcome advice to see 'old so and so', whose failings they have smiled at for years, or who is a personal friend. But it may be necessary for colleagues to insist explicitly that treatment is followed, because doctors are notorious in not even following advice they would give to their own patients (and, of course, in addiction, deciding that they will treat themselves, sometimes self-medicating in an alarming way).

The medical profession in many countries has procedures, before disciplinary procedures are instigated, of helping a sick doctor find his way to treatment. In British NHS hospitals there is the 'three wise men' arrangement, which a doctor can approach for advice about an at-risk colleague. However, there may be a stage in the evolution of a problem at which a sick doctor might be receptive to advice from outside rather than from within his own region. In the UK, a national body has been created by the Royal Colleges and the British Medical Association which any doctor seriously concerned about the effects of illness or the fitness to practise of a colleague may contact (The National Counselling and Welfare Service for Sick Doctors, tel. 071 580 3160; the Overseas Doctors Association, tel. 061 236 5594). A national adviser from within the same specialty as the patient may then contact the doctor concerned. In the US, similar hot-lines exist on a statewide basis, often run by county medical societies.

The role of the General Medical Council and accreditation boards

In the UK, the General Medical Council (GMC) controls doctors' licences to practise. Allegations of serious professional misconduct, and criminal convictions, are notified to the Council. If this raises in any way the question of ill health, or if the GMC receives information suggesting ill health, there is a screening procedure which can greatly minimize the doctor's anxiety, because legal proceedings are averted and there is no publicity. The first two years of a similar 'diversion' programme, instituted by the Californian Board of Medical Quality Assurance, were described by Gualtieri *et al.* (1983). By the second year, self-referral, albeit often instigated by peers, was a more usual route of making contact than official disciplinary action.

Procedures which allow diversion from disciplinary action enable the doctor to continue to practise, as long as he or she accepts that his behaviour is monitored. The GMC require that a supervisor acceptable to the doctor and to the GMC screener is nominated, who will obtain information and perhaps arrange to see the doctor at short notice. In addition, reports at regular intervals are required from the doctor's therapist to confirm compliance with treatment. These reports would include, for example, urine specimens in a drug abuser, markers of drinking, such as serum gamma-glutamyl transferase, and mean red cell volume in an alcoholic, or serum lithium in a manic-depressive doctor who had agreed to that form of treatment.

Assessment is the beginning of therapy

Although doctors are trained to accept psychiatric conditions as illnesses, they are often personally very sensitive to the stigma associated with psychiatric treatment. Although this should be recognized by the treating psychiatrist, it is usually best not to deviate from customary practice. Consultations should be at normal hours and at the usual clinic. The doctor's own family physician should be involved as normal. Secrecy does not facilitate good therapy.

Seeing one's condition as illness, be it anxiety state, depression, or addiction, rather than moral weakness, reduces shame and guilt: this is not incompatible with the tenet that the individual is responsible for taking appropriate steps to get well and stay well.

It is best to treat the sick doctor as an intelligent layman, and not assume that he or she knows all about psychiatric illness (although there may be an attempt to convince one of that!). The doctor's self-esteem will be at a low ebb and use of jargon and display of knowledge and experience is understandable at this point.

The interview should begin by inquiring what the individual's own worries are. It may be despair about work and inability to cope with the demands; a complaint about the marriage; the drink–driving offence which alerted the GMC; or simply anger at being told to go for a psychiatric examination. Information which is available from colleagues or employer should be discussed and consent obtained to speak to the spouse, 'to have your partner's views of how best I can help'.

If the presenting complaint would normally indicate a physical examination and blood tests, these should be done. For suspected alcohol problems, one should look for physical signs of heavy drinking, such as tremor of the fingers, mouth, and tongue; hepatomegaly; and excessive capillarization of the face or conjunctivae, or even spider naevi. The following must be checked: liver function tests, mean cell volume, and urinalysis for benzodiazepines (alcoholic doctors often self-medicate to disguise anxiety and withdrawal symptoms).

Treatment

After agreeing to treatment, the doctor may well try to terminate this prematurely. This is best averted by a warm, trusting relationship with a psychiatrist who is prepared to set limits and adhere to good clinical practice. Helpful pressure to ensure that the doctor stays in treatment and attends follow-up can be brought to bear by the doctor's colleagues or employing authority (perhaps the occupational health physician or medical personnel staff) or the GMC if it has been involved.

The doctor may need help to accept that it is appropriate to take time off work, and not to feel guilty towards colleagues or patients. Decisions about admission to hospital for treatment should be the same as for other patients. Sick doctors have been successfully treated as in-patients in the hurly-burly of NHS wards (and their subsequent comments on life in these wards can be illuminating and helpful—Anon. 1990). Similarly, decisions about physical treatments should be made as for other patients: drugs or perhaps ECT for depression, or perhaps disulfiram for relapsing alcoholism (supervised by spouse, colleague or the employer's occupational medicine department).

Psychotherapy

This may be especially relevant in two areas: the marriage, and the management of work stress. The doctor may have neglected *the emotional needs of the family*. The spouse may have years of resentment and when asked to help feels: 'Where were you when I needed support?' Addictions and depression can lead to the individual becoming solitary and self-absorbed, another cause of hardening of the partner's attitudes or demoralization.

Communication with the spouse may be helped by joint sessions in which each is encouraged to express needs and to practise being a better listener. Relearning to give and receive in the relationship will take time. As recovery begins, it is important that the patient has realistic expectations of the time it can take for the family to respond, to avoid disappointment. Playing down how hurt the family have been is part of the denial of the illness.

To have more time and energy for the family, as well as to 'cherish' him- or herself, will entail setting limits on the emotional drain of work. Therapist and patient may need to evolve guidelines for *managing stress*:

- accepting time constraints, and therefore setting boundaries to work demands;
- deciding on methods for managing difficult patients (general practitioners have patients who can be demanding, sometimes even threatening and frightening);
- finding a way to accept imposed administrative changes and only fighting those battles that are really important;
- accepting that one is expected to be 'good enough' rather than perfect;
- delegating work to others;
- not letting a 'problem' burgeon in the mind into a 'catastrophe';
- not harbouring resentments;
- rebuilding self-esteem, by taking satisfaction in successes, rather than dwelling on failures.

Monitoring progress

This is especially important when alcohol or drugs are involved. During an eight year study of the Oregon programme for helping impaired physicians on probation, the majority of whom had alcohol and/or drug problems, outcome was best (96 per cent 'improved') in those physicians who entered the programme after a procedural decision had been made that the out-patient phase of treatment would be closely monitored by random blood and urine tests. Those who had entered in the years before monitoring was instituted had not done as well (64 per cent 'improved') (Shore 1987). Close monitoring was one helpful factor proposed by Shore *et al.* (1984) when they found that alcohol and drug-abusing doctors discharged from the Mayo clinic had a better prognosis (83 per cent favourable outcome) than other discharged patients also drawn from fairly high socioeconomic groups (62 per cent favourable outcome).

If supervision is agreed to, early return to work should be encouraged, so that credibility is quickly re-established.

Relapse in addiction

One must expect and accept initial relapse in alcohol and drug abusers. The doctor's family, colleagues, and employers should be encouraged to take action as soon as relapse is noted, otherwise the addict will learn that concealment works. It is reasonable for colleagues or employer to agree a limit to the number of relapses within, say, the first year. They cannot be expected to soldier on for years with no improvement in the individual.

Mutual-help groups

These groups, of which Alcoholics Anonymous is by far the most widely distributed, can be of great help. A considerable number of doctors can testify that attendance at such meetings does not damage a doctor's reputation. In many countries it is now possible to link a drug or alcohol depend-ent physician to a national or regional 'doctor's and dentist's group'. The non-judgemental, empathic welcome the physician receives from such a group helps greatly in reducing stigma, and can help him or her accept the need for complete abstinence. That recovery can indeed be facilitated is suggested by a survey conducted amongst the 100 members of a UK regional group for alcoholic doctors which had been running for eight years (Lloyd 1990). Seven died, still in the throes of their addiction, and five had died but had been abstinent. Two had retired somewhat early. However, 76 were well and still practising. The survey only reviewed those who had continued in contact with the group for at least six months, and those who did not adhere to the group may have had a poor prognosis for other reasons. Nevertheless, Lloyd (1990) is probably correct to point out that the outcome of this sample was very considerably better than that UK sample of alcoholic doctors admitted to psychiatric hospital studied 20 years earlier by Murray (1976), amongst whom 12 per cent had died due to alcohol and only 42 per cent were in recovery, after a similar length of follow-up.

Clinical guidelines

1. Courage and compassion, not covering-up, are needed to help colleagues with emotional problems to accept responsibility and get the help they need.
2. Coercion, if supportive, can be constructive.
3. It is important to ensure that the sick doctor does not just make promises or pay lip-service to seeking help, but participates actively in treatment.
4. The dialogue with the physician–patient should be as with an intelligent layman, without assuming extra knowledge, and should commence with the problems as perceived by him or her.
5. Secrecy should be avoided.
6. The psychiatrist, non-judgemental and empathic, may need to set limits and prevent too early discontinuation of therapy.
7. Doctors with alcohol and drug problems should have their progress monitored, using objective markers, and if they agree to this could be able to return to work soon.
8. The family should be involved in treatment.
9. One should look at methods for managing work stress more efficiently.
10. Encouraging contact with a mutual-help group can be beneficial.

References

APA (American Medical Association) (1987). Results and implications of the AMA-APA physician mortality project. *Journal of the American Medical Association*, **257**, 2949–53.

Anon. (1990). View from the bottom. *Psychiatric Bulletin*, **14**, 452–4.

Arnetz, B. B., Horte, L. G., Hedberg, T., Allander, E., and Malker, H. (1987). Suicide patterns among physicians related to other academics as well as to the general population: results from a national long-term prospective study and a retrospective study. *Acta Psychiatrica Scandinavica*, **75**, 139–43.

Branthwaite, A. and Ross, A. (1988). Satisfaction and job stress in general practice. *Family Practice*, **5**, 83–93.

Collier, D. J. and Beales, I. L. P. (1989). Drinking among medical students: a questionnaire survey. *British Medical Journal*, **299**, 19–22.

Crawshaw, R., *et al.* (1980). An epidemic of suicide among physicians on probation. *Journal of the American Medical Association*, **243**, 1915–17.

Firth-Cozens, J. (1989). Stress in medical undergraduates and house officers. *British Journal of Hospital Medicine*, **41**, 161–4.

Firth-Cozens, J. (1990). Sources of stress in women junior house officers. *British Medical Journal*, **301**, 89–91.

Godlee, F. (1990). Stress in women doctors: women should not have to overcome more barriers than men. *British Medical Journal*, **301**, 76.

Gualtieri, A. C., Cosento, J. P., and Becker, J. S. (1983). The California experience with a diversion program for impaired physicians. *Journal of the American Medical Association*, **249**, 226–9.

Lloyd, G. (1990). Alcoholic doctors can recover. *British Medical Journal*, **300**, 728–30.

Lorber, J. A. and Ecker, M. (1983). Career development of female and male physicians. *Journal of Medical Education*, **58**, 447–56.

Murray, R. M. (1976). Characteristics and prognosis of alcoholic doctors. *British Medical Journal*, **2**, 1537–9.

Murray, R. M. (1977). Psychiatric illness in male doctors and controls: an analysis of Scottish hospitals in-patient data. *British Journal of Psychiatry*, **131**, 1–10.

OPCS (Office of Population Censuses and Surveys) (1986). *Decennial survey of occupational mortality*. HMSO, London.

Pitts, F., Schuller, A., and Rich, A. (1979). Suicide among U.S. female physicians, 1967–1972. *American Journal of Psychiatry*, **138**, 694–6.

Rawnsley, K. (1985). Helping the sick doctor: a new service. *British Medical Journal*, **291**, 922.

Roy, A. (1985). Suicide in doctors. *Psychiatric Clinics of North America*, **8**, 377–87.

Sakinofsky, I. (1980). Suicide in doctors and wives of doctors. *Canadian Family Physician*, **26**, 837–44.

Shore, J. H. (1987). The Oregon experience with impaired physicians: an eight year follow-up. *Journal of the American Medical Association*, **257**, 2931–4.

Shore, R. M., Martin, M., Swenson, W. M., and Niven, R. G. (1984). Prognosis of physicians treated for alcoholism and drug dependence. *Journal of the American Medical Association*, **251**, 743–6.

Strega, M. (1978). Protecting the public. *World Medicine*, 22 March, 47–8.

Surtees, P. G. and Miller, P. M. (1990). The interval general health questionnaire. *British Journal of Psychiatry*, **157**, 679–86.

Vaillant, G. E., Sobowale, N. C., and McArthur, C. (1972). Some psychological vulnerabilities of physicians. *New England Journal of Medicine*, **3**, 324–9.

Waring, E. M. (1974). Psychiatric illness in physicians: a review. *Comprehensive Psychiatry*, **15**, 519–30.

Index

DEPARTMENT OF APPLIED
SOCIAL STUDIES AND
SOCIAL RESEARCH
BARNETT HOUSE
WELLINGTON SQUARE
OXFORD OX1 2ER